Rogelio Daniel Acevedo
Maximiliano C. L. Rocca
Víctor Manuel García

Catalogue of Meteorites from South America

Springer

Rogelio Daniel Acevedo
Centro Austral de Investigaciones Científicas
Tierra del Fuego
Argentina

Víctor Manuel García
Centro Austral de Investigaciones Científicas
Tierra del Fuego
Argentina

Maximiliano C. L. Rocca
The Planetary Society
Buenos Aires
Argentina

There may be instances where the authors have been unable to trace or contact the copyright holders. If notified the publisher will be pleased to rectify any errors or omissions at the earliest opportunity.

ISSN 2191-589X ISSN 2191-5903 (electronic)
ISBN 978-3-319-01924-6 ISBN 978-3-319-01925-3 (eBook)
DOI 10.1007/978-3-319-01925-3
Springer Cham Heidelberg New York Dordrecht London

Library of Congress Control Number: 2013949406

Printed on acid-free paper

Springer is part of Springer Science+Business Media (www.springer.com)

SpringerBriefs in Earth System Sciences

South America and the Southern Hemisphere

For further volumes:
http://www.springer.com/series/10032

Acknowledgments

This brief overview of South American meteorites was partially funded by the Centro Austral de Investigaciones Científicas (CADIC) and the Consejo Nacional de Investigaciones Científicas y Tecnológicas (CONICET) of Argentina, the National Geographic/Waitt, and The Planetary Society.

We are also grateful to Jorge Rabassa for his remarks on the manuscript and to Ricardo Alonso for his valuable comments about some new and/or little-known meteorites.

Contents

Symbols and Abbreviations

S_n	Shock stage
W_n	Weathering grade
Fa	Fayalite
Fs	Ferrosilite
Wo	Wollastonite
χ	Magnetic susceptibility
L	Low iron chemical group
LL	Low iron, low metal chemical group
H	High iron chemical group
CO	Ornans chemical group
CK	Karoonda chemical group
CR	Renazzo chemical group
H_n, L_n or LL_n	Petrologic types
CO_n, CK_n or CR_n	Petrologic types
C2	Carbonaceous chondrite of petrologic type 2
A	Chemical group of irons
IVB	Chemical group of irons
Ni	Nickel
Mod	Moderate
Med	Medium

Chapter 1
Introduction

Today, it is clear that the interplanetary space between the planets of our Solar System is not a perfect void. Millions of small bodies are present there orbiting our Sun together with the large objects. If the small bodies are bigger than 100 m in diameter and they are composed of rocks and/or metals then they are called as *asteroids*. If they are integrated by ice then they are called *comets*. If they are smaller than 100 m in diameter then they are called *meteoroids*. In many cases, their orbits cross the orbit of our planet so they can get in direct contact with our atmosphere. When a meteoroid enters in our atmosphere and survives as far as to reach to the Earth's surface it is called a *meteorite*. So meteorites are small solid masses coming from the interplanetary space between the planets.

Meteorites are at present the most important and numerous sources of extraterrestrial rock or metal samples for the planetary scientific research. They may come from fragments ejected from the surfaces of asteroids in collisions between asteroids themselves, and they may also be a pieces ejected in large impact events on the surface of our Moon and in the surfaces of nearby planets like Mars and Venus.

Meteorites are very old objects (4,600 to 4,000 Ma on average) so they are important to understand the events connected with the origin of our Solar System. They also teach us about the exotic geology and geochemistry of all these extraterrestrial bodies.

The names of meteorites are founded after the places where they are fallen or found, and they are classified into three wide categories and some subcategories.

Stony meteorites are composed mostly of mafic silicate minerals. There are two types: chondrites (primitive meteorites, with chondrules) and achondrites (differentiate meteorites, without chondrules).

Stony-iron meteorites have approximately equal quantities of metal and silicates. They comprise the pallasites and mesosiderites.

Iron meteorites are especially metal. They are classified into twelve major groups depending on their relative amounts of iron, nickel, and certain trace elements.

R. D. Acevedo et al., *Catalogue of Meteorites from South America*,
SpringerBriefs in Earth System Sciences, DOI: 10.1007/978-3-319-01925-3_1,

1.1 About This Catalogue

South America is a very large continent with many different types of geological and geographical landscapes and terrains so quite large numbers of meteorites can be expected. However, the total number of meteorites from this continent only represents the 1 % of all the specimens obtained from the entire World.

One reason is that many large areas of South America are covered with dense tropical rainforests (e.g., the Brazilian Amazonas) that make extremely hard to find or recuperate meteorites. The very humid weather there also makes very difficult the preservation of samples. On the other hand, the enormous desert zones in many areas of South America (e.g., the Atacama's desert in the Pacific Ocean coast of Chile, Fig. 1.1) are prolific sources of exceptionally well-preserved specimens (Bland et al. 2000). There are about 70 % of South American meteorites came from this region and no less than 50 new specimens were reported just in the month before the close of the writing of this monograph. And only in the strewn field of El Médano have been collected more meteorites than in all the South American countries together, except Chile.

Similar geomorphological conditions of deflation surfaces can be identified in the Argentine Puna and Bolivian Altiplano, and these are also astonishingly promising terrains to find large numbers of well preserved new meteorites, but incredibly no systematic search has been performed and only a few meteorites have been discovered there.

Fig. 1.1 Atacama desert. *Credit*: Frans Swaalf

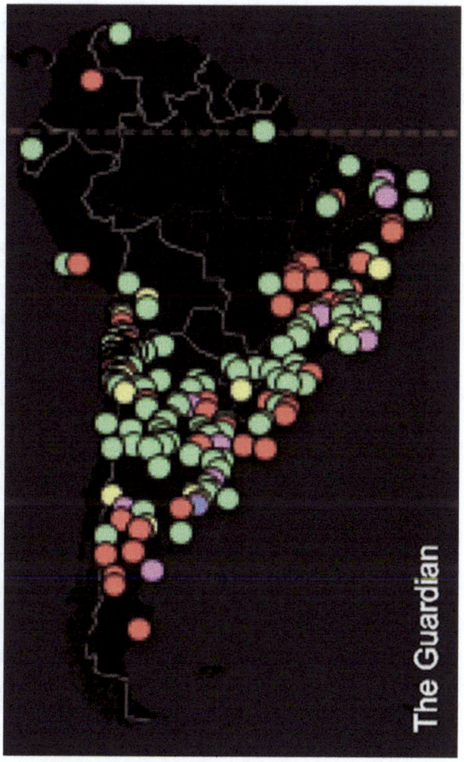

Fig. 1.2 Geographic distribution of South American meteorites. *Credit*: Simon Rogers, published in the Guardian. Mapped using CartoDB

But also most of the reported meteorite specimens from the rest of countries have been seen to fall by chance and found later or they were hit on by peasants plowing in the field.

This catalogue gives the basic information concerning each specimen like its precedence and the place where it was discovered (in geographic co-ordinates and illustrative map), the official name of the meteorite, its classification type (class, and if applicable, weathering grade and shock stage), if it was seen to fall or was accidentally found by a lucky chance, its total mass or weight, the institution who keeps it, and the most important bibliographic references about it.

Previous catalogs as Giacomelli (1969), Zucolotto (1999), Zucolotto et al. (1999), Grady (2000), Zucolotto and Antonello (2004a), Koblitz (2006), Acevedo and Rocca (2011) are also cited. However, the Database of the Meteoritical Society at http://www.lpi.usra.edu/meteor/metbul.php has been the main source of information for this new catalog.

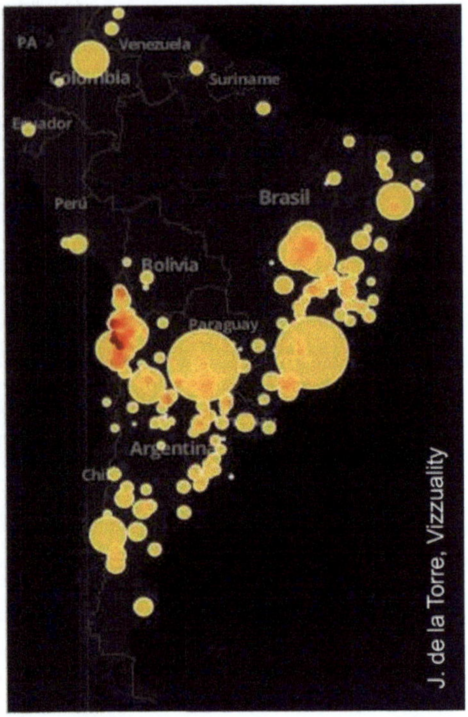

Fig. 1.3 Compared masses of South American meteorites. *Credit*: Javier de la Torre

1.2 List of Meteorites Found in South America

Figure 1.2 shows the geographic distribution of South American meteorites (falls e.g., Ecuador and finds e.g., Patagonia) and Fig. 1.3 are compared masses of the South American meteorites (displayed in proportion with the intensity of the tone and the size of the circles, that is, the largest the size of the circle and the tonality then the largest of the mass of the meteorite).

In brief, the falling of meteorites is a random event but the finding of old meteorites is focused in certain areas like, for example, in the desert of Atacama in Chile where meteorites are concentrated with the geological timescales. As it is expected, the rest of meteoroids were seen in populated and template weather areas.

So far, the number of meteorites specimens reported in this continent is 591, quantity that has been duplicated in the last year by the discovery of many new meteorite specimens from Chile.

The list displayed by country will be shown in the following chapters.

References

Acevedo RD, Rocca M (2011) Catálogo de los meteoritos hallados en territorio argentino. Revista Historia Natural Tercera Serie 1:17–34

Bland PA, Bevan RWA, Jull AJT (2000) Ancient meteorite finds and the earth's surface environment. Quatern Res 53:131–142

Giacomelli LO (1969) Guía de Meteoritos de la Argentina. Revista Museo Argentino de Ciencias Naturales, Geología, Tomo 7, No 1

Grady MM (2000) Catalogue of meteorites: with special reference to those represented in the collection of the natural history museum, 5th edn. Cambridge University Press, Edinburgh, London, 696 pp

Koblitz J (2006) MetBase, version 7.1. (CD-ROM)

Zucolotto EM (1999) Brazilian meteorites. Meteorite 5:8–11

Zucolotto ME, Antonello LL (2004a) Brazilian meteorites. In: 67th annual meeting of the meteoritical society, Rio de Janeiro (Brazil). Meteorit Planet Sci 39(Suppl):5089

Zucolotto EM, de Carvalho W, Gomes SO (1999) The Bendegó Iron. Meteorite 5(4):36–39

Chapter 2
Argentina

"The earliest Spanish explorers to enter the region that now is northern Argentina heard marvelous stories from the Indians of a large block of iron that had supposedly fallen from the sky. The place where it lay was called Pigüem Nonraltá, or, in Spanish, Campo del Cielo (Field of the Heaven). In 1576 a military expedition visited the site, returning with a few small pieces of a very large mass, which came to be known as the Mesón de Fierro (Large Table of Iron) estimated to weight about 500 Q (23 mt)" (Cassidy et al. 1965). Campo del Cielo is the flagship of the Argentine meteorites. Chaco, the main mass, is the second largest known meteorite (as a single piece) at the Earth's surface.

Lists of all known Argentine meteorites were published by Giacomelli (1969), and Acevedo and Rocca (2008, 2010, 2011). In addition, data for all specimens hitherto identified (Fig. 2.1) are included here.

Argentine meteorites are protected by National Law N° 26,306. Any meteorite found or fall in Argentina is a cultural treasure of the Argentine Nation and its commerce is forbidden.

Achiras. 33° 10′ S, 64° 57′ W. Río Cuarto, Córdoba. Stone, olivine-hypersthene chondrite, (L_6). Fell on 1902, 780 g were recovered. Museo Provincial de Ciencias Naturales "Bartolomé Mitre", Córdoba (Argentina), Smithsonian Institution collections (USA), Natural History Museum (UK). Olsacher (1951a).

Agua Blanca. 28° 55′ S, 66° 57′ W. Pinchas, Castro Barros, La Rioja. Iron, octahedrite, (IIIAB). Collected in 1938, 49 kg obtained. Museo "Inca Huasi", La Rioja (Argentina), Natural History Museum. Herrero Ducloux and Loyarte (1939).

Aguada. 31° 36′ S, 65° 14′ W. Pocho, Córdoba. Stone, olivine-hypersthene chondrite, (L_6). Fell on September, 1930, 1.62 kg were recovered. Museo de Mineralogía y Geología, Facultad de Ciencias Exactas, Físicas y Naturales, Córdoba, Smithsonian Institution collections, Natural History Museum. Olsacher (1951b). Fig. 2.2.

Aguas Calientes. 25° 30′ S, 68° 24′ W. Catamarca. Stone, olivine-bronzite chondrite, (H). Find, 1971, 257 g.

Águila Blanca. 30° 52′ S, 64° 33′ W. Río Dolores, Punilla, Córdoba. Stone, olivine-hypersthene chondrite, (L). Date of find: 1920, 1.44 kg. Museo Argentino

R. D. Acevedo et al., *Catalogue of Meteorites from South America*, SpringerBriefs in Earth System Sciences, DOI: 10.1007/978-3-319-01925-3_2,

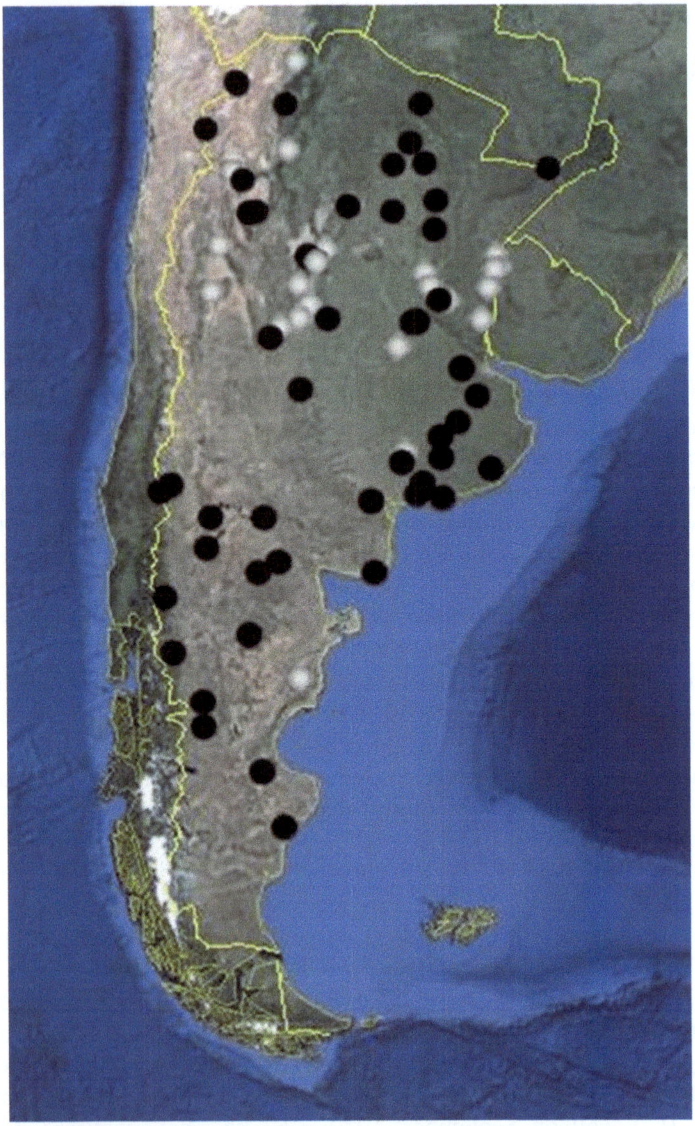

Fig. 2.1 Argentine meteorites. Modified from © 2013 Inav/Geosistemas SRL

de Ciencias Naturales "Bernardino Rivadavia", Buenos Aires. (Herrero Ducloux 1939).

Árbol Solo. 33° S, 66° W. Sococora, San Luis. Stone, olivine-bronzite chondrite, (H_5). Fell on Septiember 11, 1954, 809 g. Universidad Nacional de Cuyo, Mendoza (Argentina). The Permanent Commission on Meteorites of the International Geological Congress (1964b); Giacomelli (1969). Fig. 2.3.

Fig. 2.2 Aguada. *Credit*: Eduardo Jawerbaum

Fig. 2.3 Árbol Solo. *Credit*: Eduardo Jawerbaum

Fig. 2.4 Arroyo Aguiar. *Credit*: Matteo Chinellato

Arroyo Aguiar. 31° 25′ S, 60° 40′ W. Santa Fe. Stone, olivine-bronzite chondrite, (H_5). Fall, 1950, 7.45 kg. Museo de la Facultad de Ingeniería Química, Santa Fe (Argentina), Smithsonian Institution collections, Natural History Museum. Benet (1961). The Permanent Commission on Meteorites of the International Geological Congress 1962b). Fig. 2.4.

Balcarce. 37° 52′ S, 58° 15′ W. Cerro Amarante, Balcarce. Buenos Aires Province. Stone, olivine-bronzite chondrite, (H_4). Found on June, 2000, 2.28 kg. Turone (2001). (It has not been officially classified). Fig. 2.5.

Belville. 32° 20′ S, 64° 52′ W. Unión, Córdoba. Stone, chondrite. Fell on December, 1937. Unknown mass.

Berduc. 31° 55′ S, 58° 20′ W. Entre Ríos. Stone, chondrite, (L_6), S_4 W_0. Seen falling on April 6, 2008. Several kilograms were drawn from Argentina. Some

Fig. 2.5 Balcarce. *Credit*: Eduardo Jawerbaum

Fig. 2.6 Berduc. *Credit*: Michael Farmer

pieces have been treasured by Asociación Entrerriana de Astronomía. MNCNA-AS (Weisberg et al. 2009; Varela et al. 2010). Fig. 2.6.

Cacharí. 36° 24′ S, 59° 30′ W. Cacharí, Azul, Buenos Aires Province. Achondrite, eucrite. Collected in 1916, 23.5 kg. Museo de La Plata (Argentina), Smithsonian Institution collections, Natural History Museum. (Herrero Ducloux 1929; Abdu et al. 2005a, b). Fig. 2.7.

Campo de Pucara. 27° 40′ S, 67° 07′ W. Andalgalá, Catamarca. Iron, hexahedrite. Find, 1879, 4 kg. Specimens can be found in the collections of many European museums. Giacomelli (1969).

Campo del Cielo. 27° 30′ S, 61° 42′ W. Chaco, Iron, octahedrite, (IAB). Findings from 1576. Several tons recovered. The meteorite field consists, at least, of 20 meteorite craters with an age of about 4000 years.

A lot of papers can be consulted on the subject (Nágera 1926; Fossa Mancini 1948; Cassidy 1967, 1968, 1971; Villar 1968; Cassidy and Renard 1996; Cassidy et al. 1965; Acevedo and Rocca 2005; Cabanillas and Palacios 2006; Wright et al. 2007).

The area is composed of sandy-clay sediments of Quaternary-Recent age. The impactor was an Iron-Nickel Apollo-type asteroid (octahedrite meteorite type IA) and plenty of meteorite specimens survived the impact. Impactor's diameter is estimated to 5–20 m. Its diameter pre-atmospheric was estimated (based on

Fig. 2.7 Cacharí. *Credit*: Museo de La Plata

measurements of cosmogenic radioisotopes) such as about 3 m and a mass of about 840 t (Liberman et al. 2002). Touring a solar orbit calculated some years ago (Renard and Cassidy 1971), the impactor came from the SW and entered into the Earth's atmosphere in a low angle of about 9°. As a consequence, the asteroid broke in many pieces before creating the craters. The first meteorite specimens were discovered during the time of the Spanish colonization. Craters and meteorite fragments are widespread in an oval area of 18.5 × 3 km (SW-NE), thus Campo del Cielo is one of the largest meteorite crater fields known in the world.

Crater n° 3, called "Laguna Negra" is the largest with a diameter of 115 m. Inside crater n° 10, called "Gómez", (diameter about 25 m), a huge meteorite specimen called "El Chaco", of 37.4 tons, was found in 1980. Inside crater n° 9, called "La Perdida" (diameter: 25 × 35 m.) several meteorite pieces were discovered weighing in total about 5200 kg. Other best known crater is Rubin de Celis with an enormous schreibersite-bearing metallic block (Acevedo et al. 2002).

The following is a list of large meteorite specimens (more than 200 kg) from this area cited by Rocca (2006). Information is given in the name of the meteorite its weight, and date of discovery and its current location:

(1) El Abispón. 460 kg. 1936. Museo Argentino de Ciencias Naturales (MACN), Buenos Aires city. The Permanent Commission on Meteorites of the International Geological Congress (1962a).
(2) El Chaco. 37.4 tons. 1980. Gancedo, Chaco, Argentina. This specimen is the second heaviest meteorite known in the World. Fig. 2.8.
(3) El Mataco. 998 kg. 1937. Museo Provincial, Rosario, Santa Fe, Argentina. The Permanent Commission on Meteorites of the International Geological Congress (1962a).

Fig. 2.8 El Chaco. *Credit*: L.M. Villar

(4) El Mocoví. 732 kg. 1925. MACN, Buenos Aires city.
(5) El Patio. 350 kg. Found before 1960. Estancia el Taco, Chaco, Argentina. The Permanent Commission on Meteorites of the International Geological Congress (1965).
(6) El Taco. 1998 kg. 1962. Main mass at National Museum of Natural History, Smithsonian Institution collections. 600 kg. at Planetarium of Buenos Aires city. The Permanent Commission on Meteorites of the International Geological Congress (1965). Fig. 2.9.
(7) El Toba. 4,210 kg. 1923. MACN, Buenos Aires city.
(8) El Toconoté. 850 kg. 1931. Planetarium, Buenos Aires city. The Permanent Commission on Meteorites of the International Geological Congress (1962a).
(9) La Perdida (1). 1,625 kg. 1965. Planetarium, Buenos Aires city.
(10) La Perdida (2). 3,370 kg. 1965. Still in the crater.
(11) Mesón de Fierro. 15 tons. 1576. Lost.
(12) Runa Pocito. 750 kg. 1803. British Museum, London.
(13) No Name. 10 tons. 1997. Near its finding site, Chaco.
(14) La Sorpresa. 7/10 tons. 2005. Still in the crater.
And others recently discovered such as Adolfo 9,760 kg, Quimili 8,000 kg, Santiagueño 7,850 kg and Carmen Sosa 5,680 kg.

Caperr. 45° 17′ S, 70° 29′ W. Río Senguerr, Chubut. Iron, octahedrite, (IIIA or IIIAB). Collected in 1869, is a single stone of 114 kg. Museo de La Plata, Smithsonian Institution collections, Natural History Museum. Fletcher (1899), Turone (2002). Fig. 2.10.

Capilla del Monte. 30° 53′ S, 64° 33′ W. Córdoba. Stone, olivine-bronzite chondrite, (H_6). Fell on 1934. Fragment of approximately 1 kg is at the Museum of the Dirección Nacional de Geología y Minería in Buenos Aires, Smithsonian

Fig. 2.9 El Taco. *Credit*: Unknown

Fig. 2.10 Caperr. *Credit*: Museo de La Plata

Institution collections and Natural History Museum. The Permanent Commission on Meteorites of the International Geological Congress (1965), Giacomelli (1969). Fig. 2.11.

Casilda. 33° 06′ S, 61° 08′ W. Casilda, Santa Fe. Stone, olivine-bronzite chondrite, (H_5). Find, 1937, 5.25 kg. Universidad Nacional de Tucumán (Argentina), Smithsonian Institution collections, Natural History Museum. Toselli and Brodtkorb (1973). Clarke Jr. (1974). Fig. 2.12.

Cerro Mesa. 46° 51′ S, 68° 08′ W. Santa Cruz. Stone, chondrite (L or LL_6). Found in 2006, 10.5 kg. (It has not been officially classified). Fig. 2.13.

Chajarí. 30° 47′ S, 58° 03′ W. Federación, Entre Ríos. Stone, olivine-hypersthene chondrite, (L_5). Seen falling on November 29, 1933, 18.3 kg were recovered. The Permanent Commission on Meteorites of the International Geological Congress (1961), Teruggi (1968), Giacomelli (1969). Fig. 2.14.

Fig. 2.11 Capilla del Monte. *Credit*: Eduardo Jawerbaum

Fig. 2.12 Casilda. *Credit*: Eduardo Jawerbaum

Fig. 2.13 Cerro Mesa. *Credit*: McCartney collection

Claromecó. 38° 48′ 16″ S, 60° 07′ 23″ W. Claromecó, Buenos Aires Province. Stone, Ordinary chondrite, (L), S_5 W_2. Olivine $Fo_{75.0}$ $Fa_{25.0}$, Pyroxene $En_{80.21}$ $Fs_{18.45}$ $Wo_{1.34}$. Two reddish brown meteoritic stones of 13 kg each were found in 1963 leaning against a windmill in a farm. Museo Aníbal Paz (Argentina). Acevedo and Subías (2012). It is not yet officially classified. Fig. 2.15.

Coronel Arnold. 33° 04′ S, 61° W. San Lorenzo, Santa Fe. Stone, olivine-hypersthene chondrite, (L). Collected in 1962, 450 g were recovered.

Cruz del Eje. 30° 45′ S, 64° 47′ W. Córdoba. Iron, coarsest octahedrite, (IAB complex). Ni = 68.8 mg/g. This iron of 14 kg was found in 1971. Cascadia Meteorite Laboratory, Portland (USA), Lakehead University, Ontario (Canada). Weisberg et al. (2010).

Dadin. 38° 55′ S, 69° 12′ W. Plaza Huincul, Neuquén. Iron, octahedrite. Ni = 6.7 wt %. Find, 1949, 37.3 kg. Museo de la Plata. Herrero Ducloux (1949) (Fig. 2.16).

Fig. 2.14 Chajarí. *Credit*: Eduardo Jawerbaum

Fig. 2.15 Claromecó. *Credit*: Fermín Massigoge

Fig. 2.16 Dadin. *Credit*: Museo de La Plata

Deán Funes. 30° 26′ S, 64° 12′ W. Deán Funes, Córdoba. Stone, ordinary chondrite, (H_5), S_2 W_1. $Fa_{19.6}$ $Fs_{17.4}$ $Wo_{1.3}$. Fall in 1977, 9.26 kg. Grossman (1998).

Distrito Quebracho. 31° 53′ S, 60° 28′ W. Paraná, Entre Ríos Stone, olivine-bronzite chondrite, (H). Fell on March 13, 1957, 400 g were recovered. Museo de Entre Ríos (Argentina), Natural History Museum. Gordillo (1959), The Permanent Commission on Meteorites of the International Geological Congress (1962a).

D'Orbigny. 37° 40′ S, 61° 39′ W. D'Orbigny, Coronel Suárez, Buenos Aires Province. Achondrite, angrite. A single piece of 16.55 kg was found in July, 1979, whereas peasants were plowing a field. Grossman and Zipfel (2001), Kurat et al.

Fig. 2.17 D'Orbigny.
Credit: Jeff Kuyken

Fig. 2.18 El Mapuche.
Credit: Unknown

(2001, 2003, 2005), Varela et al. (2001, 2003a, b, 2005), Eugster et al. (2002), Jotter et al. (2002, 2003), Mittlefehldt et al. (2002), Kubny et al. (2003), Jagoutz et al. (2003), Scorzelli et al. (2004), Abdu (2005a, b). Fig. 2.17.

El Aybal. 24° 52′ S, 65° 29′ W. International airport, Salta. Meteorite with wüstite and silicides (gupeiite, xifengite, fersilicite and ferdisilicite). Was found in 1998, 30 g. Morello and Anesa (2000). (Not officially listed yet in the Meteoritical Bulletin Database).

El Mapuche. 37° 52′ S, 71° 05′ W. Caviahue, Neuquén. Stone, chondrite. Find, 1963. Weight unknown. The meteorite belonged to the collection of the Museum. Prof. Dr. John A. Olsacher (Zapala, Neuquén) while José Ignacio Garate Zubillaga was the curator. C.f Ñorquín-co meteorite (found 18 years before 45 km away to the northeast). (It has not been officially classified). See Fig. 2.18.

El Perdido. 38° 41′ S, 61° 06′ W. Buenos Aires Province. Stone, olivine-bronzite chondrite, (H_5). Found in 1905, possibly belonging to the meteorite fall of Indio Rico (Herrero Ducloux 1911), 30.25 kg were recovered. Museo de La Plata, Smithsonian Institution collections, Natural History Museum. Fossa Mancini (1947a). Fig. 2.19.

Fig. 2.19 El Perdido. *Credit*: Eduardo Jawerbaum

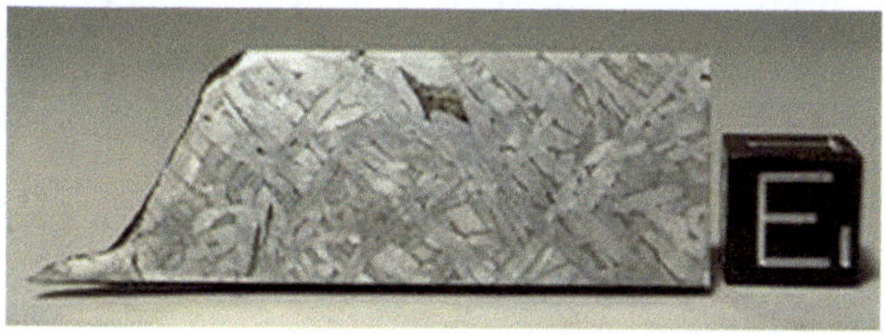

Fig. 2.20 El Sampal. *Credit*: Woreczko Jan & Wadi

El Sampal. 44° 32′ S, 70° 22′ W. Nueva Lubecka, Chubut. Iron, octahedrite, (IIIAB). Find, 1973, 142 kg. Smithsonian Institution collections, Natural History Museum. Lewis et al. (1974), Clarke Jr. (1975). Fig. 2.20.

El Simbolar. 30° 38′ S, 64° 53′ W. Cruz del Eje, Córdoba. Iron, octahedrite. Find, 1938, 40 kg. Museo Provincial de Ciencias Naturales "Bartolomé Mitre", Natural History Museum. Olsacher (1939), The American Mineralogist (1955) and reference cited therein, The Permanent Commission on Meteorites of the International Geological Congress (1958a).

El Timbú. 33° 07′ S, 60° 58′ W. San Lorenzo, Santa Fe. Iron. Found in 1942, 500 kg. Dirección de Geologia y Mineria, Buenos Aires (Argentina). The Permanent Commission on Meteorites of the International Geological Congress (1962a), Giacomelli (1969).

Esquel. 42° 54′ S, 71° 20′ W. Esquel, Chubut. Stony-Iron, pallasite. One specimen weight about 1,500 kg of the most beautiful meteorite of the world was found in 1951. Thanks to its extraordinary beauty is a meteorite found in many private collections. Smithsonian Institution collections, Natural History Museum. Giacomelli (1962, 1969), The Permanent Commission on Meteorites of the International Geological Congress (1964a). Fig. 2.21.

Fortuna. 35° 08′ S, 65° 22′ W. San Luis. Achondrite, winonaite, $W_{0/1}$ S_2. $Fa_{4.3\text{-}5.1}$, $Fs_{5.5\text{-}6.7}$ $Wo_{1.1\text{-}1.7}$ y $Fs_{2.4\text{-}2.9}$ $Wo_{45.6\text{-}46.5}$, $An_{14.5\text{-}16}$. Fe-Ni and troilite. ^{TM}170 = +1.61 ‰, ^{TM}18O = +4.68 ‰. Find, 1998, 312 g. Russell et al. (2003) (Fig. 2.22).

Gan Gan. 42° 40′ S, 68° 05′ W. Gan-Gan, Chubut. Iron, octahedrite, (IVA). Bulk metal Ni = 9.12 w%. Find, 1984, 83 kg. Institute of Geophysics & Planetary Physics, University of Los Angeles, California (IGPP-UCLA) (USA), Natural History Museum. Grossman (1999). Fig. 2.23.

Fig. 2.21 Esquel. *Credit*: Robert Haag

Fig. 2.22 Fortuna. *Credit*: Eduardo Jawerbaum

Garabato. 28° 52′ S, 60° 12′ W. Santa Fe. Stone, ordinary chondrite, (H_5). Fa_{19}, Fs_{17}. 25 % Fe-Ni and troilite. Find, 1995, 160 kg. Grossman (1996). Fig. 2.24.

Gualeguaychú. 33° 00′ S, 58° 37′ W. La Constancia, Entre Ríos. Stone, olivine-bronzite chondrite, (H). Fell on October, 1932, 22 kg were raised. Museo Argentino de Ciencias Naturales, Buenos Aires, Smithsonian Institution collections, Natural History Museum. Herrero Ducloux (1940).

Hinojal. 32° 22′ S, 60° 09′ W. Victoria, Entre Ríos. Stone, olivine-hypersthene chondrite, (L_6). Found in 1927 while plowing a field. 50 kg. Smithsonian Institution collections. The Permanent Commission on Meteorites of the International Geological Congress (1964a), Giacomelli (1969). Fig. 2.25.

Hinojo. 36° 52′ S, 60° 10′ W. Hinojo, Olavarría, Buenos Aires Province. Stone, olivine-bronzite chondrite, (H). Find, 1928, 1,155 kg. Museo de La Plata. Herrero Ducloux (1928a).

Fig. 2.23 Gan Gan. *Credit*: Woreczko Jan & Wadi

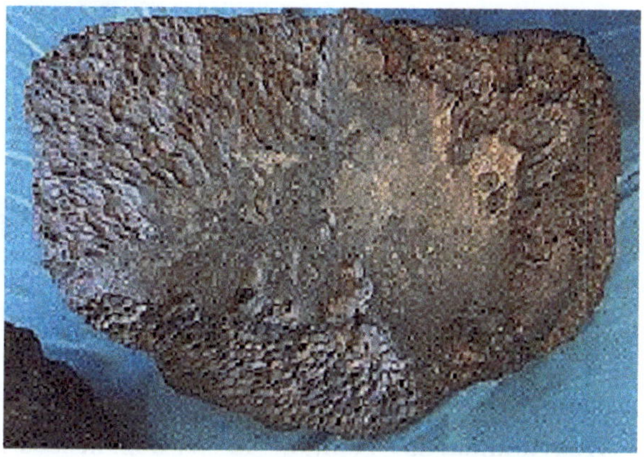

Fig. 2.24 Garabato. *Credit*: Martin Horejsi

Fig. 2.25 Hinojal. *Credit*: Mike Bandli

Fig. 2.26 Huaytiquina. *Credit*: Eduardo Jawerbaum

Fig. 2.27 Indio Rico. *Credit*: R. Kempton

Huaytiquina. 23° 44′ S, 67° 14′ W. Iron, hexaedrite IIAB. Find, 1998, 19.6 kg. Huaytiquina is a Chilean locality, next to the Argentine border (It has not been officially classified) (Fig. 2.26).

Indio Rico. 38° 20′ S, 60° 53′ W. Indio Rico, Coronel Dorrego, Buenos Aires Province. Stone, olivine-bronzite chondrite, (H_6), *C.f.* El Perdido chondrite. Find, 1887, 15 kg. Museo de La Plata, Smithsonian Institution collections. Fossa Mancini (1947b). Fig. 2.27.

Isthilart. 31°' 11′ S, 57° 57′ W. Federación, Entre Ríos Stone, olivine-bronzite chondrite, (H_5). Seen to fall on November 12, 1928, 3.05 kg were recovered. Museo Argentino de Ciencias Naturales, Buenos Aires, Smithsonian Institution collections, Natural History Museum. Herrero Ducloux y Pastore (1930).

Juárez. 37° 33′ S, 60° 09′ W. Buenos Aires Province. Stone, olivine-hypersthene chondrite, (L_6). Find, 1938, 6.1 kg. Museo Argentino de Ciencias Naturales, Buenos Aires, Smithsonian Institution collections, Natural History Museum. The

Fig. 2.28 La Colina. *Credit*: Woreczko Jan & Wadi

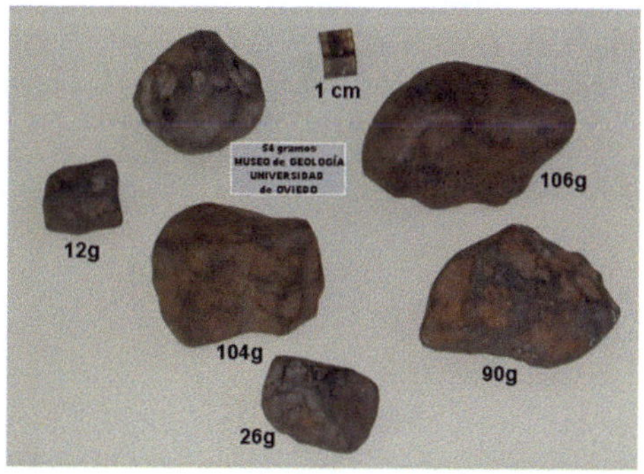

Fig. 2.29 La Criolla. *Credit*: R.D. Acevedo

Permanent Commission on Meteorites of the International Geological Congress (1962a), Giacomelli (1969).

La Colina. 37° 20′ S, 61° 32′ W. General Lamadrid, Buenos Aires Province. Stone, olivine-bronzite chondrite, (H_5). Seen falling on March 19, 1924, 2 kg were recovered. Museo Argentino de Ciencias Naturales, Buenos Aires, Smithsonian Institution collections, Natural History Museum. Herrero Ducloux (1925). Pastore (1925a). Fig. 2.28.

La Criolla. 31° 16′ S, 58° 06′ W. La Criolla, Entre Ríos Stone, olivine-hypersthene chondrite, (L_6). Fell on January 6, 1985. Meteoritic shower. 35 kg were recovered. Museo de Geología de la Universidad de Oviedo (España), Smithsonian Institution collections, Natural History Museum. SEAN Bulletin (1985), Graham (1986), Acevedo et al. (2005), Rocca (2005). Fig. 2.29.

Laguna Manantiales. 48° 35′ S, 67° 25′ W. Deseado, Santa Cruz. Iron, (IIIAB?). Find, 1945, 92 kg. Smithsonian Institution collections, Natural History Museum. The Permanent Commission on Meteorites of the International Geological Congress (1962a), Giacomelli (1969). Fig. 2.30.

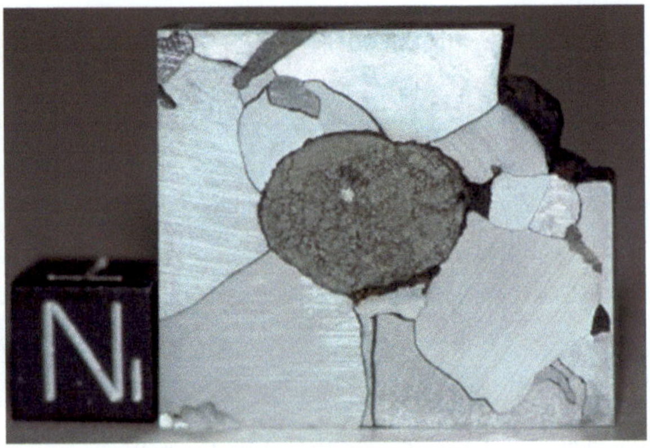

Fig. 2.30 Laguna Manantiales. *Credit*: Anne Black

Fig. 2.31 Los Cerrillos. *Credit*: Michael S. Scherman

Los Cerrillos. 28° 58′ S, 63° 20′ W. Santiago del Estero. Stone, ordinary chondrite, (H_4), S_2 W_2. $Fa_{17.3}$, $Fs_{15.1}$ $Wo_{0.9}$. Find, 2006,1 kg. Weisberg et al. (2009). Fig. 2.31.

Luján. 34° 40′ S, 59° 22′ W. Luján, Buenos Aires Province. Iron, octahedrite? weighing 50 g was found before 1878 by the paleontologist Florentino Ameghino at the depth of 6 meters in a ravine of Río Luján, 8 km southwest of Luján city, near Jáuregui town. This meteorite fell in prehistoric times (20.000 to 50.000 years ago, Late Pleistocene). Main mass is deposited in the Museo de La Plata. Ameghino (1914).

Malotas. 28° 56′ S, 63° 14′ W. Salavina, Santiago del Estero. Stone, olivine-bronzite chondrite, (H_5). Seen falling on June 22, 1931. Meteoritic shower. Hundreds of fragments recovered. Unknown mass. Smithsonian Institution collections, Natural History Museum. Olsacher (1931). Fig. 2.32.

Fig. 2.32 Malotas. *Credit*: Woreczko Jan & Wadi

Medanitos. 27° 15′ S, 67° 30′ W. Tinogasta, Catamarca. Achondrite, eucrite (brecciated). Seen falling on July 14, 1953, 31 g were rescued. Specola Vaticana (Vaticano City State), Natural History Museum. Salpeter (1957), The Permanent Commission on Meteorites of the International Geological Congress (1958b), Giacomelli (1969).

Mercedes. 34° 40′ S, 59° 20′ W. Mercedes, Buenos Aires. Stone, olivine-bronzite chondrite, (H_5), W_3. $Fa_{18.9}$, $Fs_{17.6}$ $Wo_{1.2}$. Find, 1994, 3.3 kg (26 pieces). Connolly Jr. et al. (2006). Fig. 2.33.

Muelle Viejo. 41° 11′ S, 71° 23′ W. Lago Nahuel Huapi, San Carlos de Bariloche, Río Negro. Iron, octaedrite?. Found at the depth of 8.5 meters in an excavation at the shores of Lake Nahuel Huapi in 1961, 130 g were recovered. Sección Geología, Museo de la Patagonia "Francisco P. Moreno", Bariloche (Argentina). Rivas et al. (1976). (It has not been officially classified).

Nahuel Niyeu. 40° 32′ S, 66° 38′ W. Río Negro. Stone, ordinary chondrite, (H_5), S_2 $W_{2/3}$. Find, 2005, 10.54 kg. (Weisberg et al., 2010).

Nicolás Levalle. 38° 51′ 01″ S 62° 52′ 44″ W. Nicolás Levalle, Buenos Aires Province. Stone, ordinary chondrite, (L_5), S_2 W_5. Find, 1956, 60 kg. Main mass, Sociedad Meteorítica Argentina, IGPP-UCLA. Rönick et al. (2011), Garvie (2012).

Nogoyá. 32° 22′ S, 59° 50′ W. Entre Ríos. Stone, carbonaceous chondrite, (CM_2). Fell on June 30, 1879, 4 kg. Main mass in Museo Argentino de Ciencias Naturales, Buenos Aires, Smithsonian Institution collections, Natural History Museum. Herrero Ducloux (1914), Nazarov et al. (1998). Fig. 2.34.

Ñorquin-Có. 337° 43′ S, 70° 37′ W. Neuquén. Iron, octahedrite, (IIIAB). Found in 1945, 19.25 kg. Museo de La Plata. Herrero Ducloux (1945a). Fig. 2.35.

Palca de Aparzo. 23° 07′ S, 65° 06′ W. Jujuy. Stone, ordinary chondrite, (L_5). Fa_{25}, Fs_{21}. Seen falling on September 14, 1988, 1,439 kg were rescued. Wlotzka (1994).

Fig. 2.33 Mercedes. *Credit*: Woreczko Jan & Wadi

Fig. 2.34 Nogoyá. *Credit*: www.meteorite.fr

Fig. 2.35 Ñorquin-Có. *Credit*: Mike Bandli

Pampa del Infierno. 26° 41′ S, 61° 05′ W. Avia Terai, Chaco. Stone, olivine-hypersthene chondrite, (L_6). Find, 1895, 896 g. Museo Argentino de Ciencias Naturales, Buenos Aires, Smithsonian Institution collections, Natural History Museum. Herrero Ducloux (1928b).

Pitino. 27° 28′ S, 60° 35′ W. Chaco. Stone, olivine-bronzite chondrite, (H_5), S_4 W_2. $Fa_{15.8}$, $Fs_{14.9}$ $Wo_{1.4}$. Found in 2002, 1,667 g. (Russell et al., 2003). Fig. 2.36.

Puerta de Arauco. 28° 53′ S, 66° 40′ W. La Rioja. Iron, octahedrite, (IVA?). Find, 1904, 1.5 kg. Museo de La Plata (Herrero Ducloux, 1908) (Fig. 2.37).

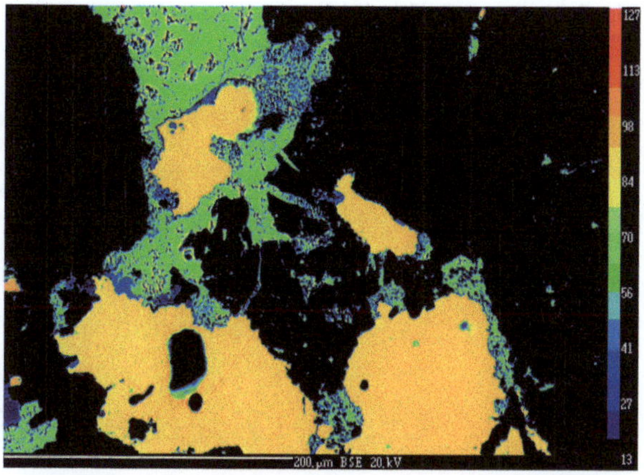

Fig. 2.36 Pitino. SEM image. *Credit*: Mariluz Valín-Alberdi

Fig. 2.37 Puerta de Arauco. *Credit*: Museo de La Plata

Raco. 26° 40′ S, 65° 27′ W. Tafí del Valle, Tucumán. Stone, olivine-bronzite chondrite, (H_5). Fell on November 17, 1957, 5 kg. Instituto Miguel Lillo, Universidad Nacional de Tucumán (Argentina), Smithsonian Institution collections, Natural History Museum. The Permanent Commission on Meteorites of the International Geological Congress (1961), Giacomelli (1969). Fig. 2.38.

Renca. 32° 45′ S, 65° 17′ W. San Luis. Stone, olivine-hypersthene chondrite, (L_5). Seen falling on June 20, 1925, 300 g were saved. Museo Argentino de Ciencias Naturales, Buenos Aires, Smithsonian Institution collections, Natural History Museum. Herrero Ducloux and Pastore (1929).

Rincón. 23° 52′ 15′ S, 67° 10′ 35′ W. Salta. Stone, ordinary chondrite, (L_6), S_4 W_2. $Fa_{25.7}$, $Fs_{22.5}$. Ringwoodite-bearing. Collected in 1995, 294.4 g were recovered. Wlotzka (1995), Stelzner et al. (1997). Fig. 2.39.

Río Cuarto 001.32° 52′ S, 64° 13′ W. Córdoba. Achondrite, eucrite. Plagioclase $An_{89.7–96.6}$, ortopyroxene $Fs_{37.2–77.1}$ $Wo_{0.3–4.8}$, pigeonite $Fs_{62.6–85.8}$ $Wo_{5.3–16.6}$, and augite $Fs_{35.4–72.6}$ $Wo_{21.3–43.5}$. Found 62.7 g within an oblong

Fig. 2.38 Raco. *Credit*: Eduardo Jawerbaum

Fig. 2.39 Rincón. *Credit*: Klaus Heide

depression of 4 x 0.6 km. University of Río Cuarto (Argentina), University of Chicago (USA), Imperial College London (UK). Connolly et al. (2007), Levine et al. (2007).

Río Limay. 39° 51′ S, 69° 29′ W. Quiñihuau, Río Negro. Stone, ordinary chondrite, (L_5). Fa_{23} Fs_{19}, plagioclase An_{15}. Find, 1995, 280 kg. Smithsonian Institution collections. Grossman (1996). Fig. 2.40.

San Borjita. 27° 33′ 31′ S, 56° 08′ W. Corrientes. Stone, ordinary chondrite, (L_4). $Fa_{24.4}$, $En_{19.8}$ $Wo_{0.8}$. S_3 W_2. Found in 1983, 12.3 kg. Smithsonian Institution collections. Grossman (1998).

San Carlos. 35° 32′ S, 58° 46′ W. San Miguel del Monte, Buenos Aires Province. Stone, olivine-bronzite chondrite, (H_4). Find, 1942, 3.6 kg. Dissapeared from the Asociación Amigos de la Astronomía de Ciudad de Buenos Aires (Argentina). Herrero Ducloux (1942). Smithsonian Institution collections, Natural History Museum. Fig. 2.41.

San Luis. 33° 20′ S, 66° 23′ W. San Luis. Stone, olivine-bronzite chondrite, (H). Find, 1964, mass unknown. Smithsonian Institution collections.

Fig. 2.40 Río Limay. *Credit*: Woreczko Jan & Wadi

Fig. 2.41 San Carlos. *Credit*: Collection Jean Michel Masson

Santa Isabel. 33° 54′ S, 61° 42′ W. Santa Fe. Stone, olivine-hypersthene chondrite, (L_6). Seen falling on November, 1924, 5.5 kg were collected. Museo Argentino de Ciencias Naturales, Buenos Aires, Natural History Museum. Herrero Ducloux (1926), Pastore (1925b).

Santa Lucía. 31° 32′ S, 68° 29′ W. Derqui and Edison streets, Villa Manuelita, Departamento de Santa Lucía, San Juan Province. Stone, ordinary chondrite, (L_6), W_0. Olivine $Fa_{24.4}$ and pyroxenes $Fs_{20.7}$ $Wo_{1.5}$ (Varela et al. 2010). Seen falling on January 23, 2008, 1.9 kg were rescued. Monomictic breccia. CASLEO, San Juan (Argentina) (354 g) and Museo de La Plata (1,526 g). (Weisberg et al., 2010). Fig. 2.42.

Sierra Colorada. 40° 48′ S, 67° 29′ W. Río Negro. Stone, ordinary chondrite, (L_5). $Fa_{24\text{-}26}$, Fs_{26}. Find, 1995, 71.3 kg. Smithsonian Institution collections. Grossman (1997). Fig. 2.43.

Talampaya. 29° 55′ S, 68° 05′ W. La Rioja. Achondrite, eucrite. $En_{58.6\text{-}60.0}$ $Wo_{1.2\text{-}1.6}$ $En_{40.5}$ $Wo_{45.7}$, plagioclase $An_{89\text{-}95}$, chromite with 1.26 % MgO, very low concentration of incompatible trace elements, and Cr 3,400 ppm, typical of cumulate eucrites. Fall, 1995, 1,421 kg. Natural History Museum. Grossman (1999) Fig. 2.44.

Fig. 2.42 Santa Lucía. *Credit*: Diario de Cuyo

Fig. 2.43 Sierra Colorada. *Credit*: Woreczko Jan & Wadi

Fig. 2.44 Talampaya. *Credit*: courtesy of Heritage Auctions, www.HA.com

Fig. 2.45 Vera. *Credit*: Eduardo Jawerbaum

Fig. 2.46 Viedma. *Credit*: Woreczko Jan & Wadi

Tostado. 29° 14′ S, 61° 46′ W. Santa Fe. Stone, olivine-bronzite chondrite, (H). Find, 1945, 22 kg. Museo Argentino de Ciencias Naturales, Buenos Aires, Smithsonian Institution collections, Natural History Museum. Herrero Ducloux (1945b).

Tres Estacas. 25° 09′ S, 60° 40′ W. Río Teuco, Chaco. Stone, ordinary chondrite. Find, 1968, 160 kg. Villar (1968).

Uzcudún. 44° 07′ S, 66° 09′ W. Ameghino, Chubut. Stone, olivine-hypersthene chondrite, (L). Fell on April 16, 1948, 20 kg. Museo Histórico Municipal de Bahía Blanca (Argentina). De Serralonga (1969), Wlotzka (1992).

Vera. 29° 55′ S, 60° 17′ W. Vera, Santa Fe. Stone, olivine-hypersthene chondrite, (L or LL4). Find, 1941, 80 kg. Smithsonian Institution collections, Natural History Museum. The Permanent Commission on meteorites of the International Geological Congress (1961), Carnevali (1953). Fig. 2.45.

Viedma. 41° 04′ S, 62° 51′ W. Río Negro. Stone, olivine-hypersthene chondrite, (L_5). $Fa_{24.9}$ $Fs_{21.1}$. Found in 2003, 6.9 kg. Russell et al. (2004). Fig. 2.46.

Fig. 2.47 Villa Regina. *Credit*: Woreczko Jan & Wadi

Villa Regina. 39° 06′ S, 67° 04′ W. Río Negro. Iron, (IIIAB). Ni = 79.3 mg/g. Find, 2005, 5.03 kg. Published in Meteoritical Bulletin, N° 100, MAPS 46, in preparation (2013). Fig. 2.47.

References

Abdu Y, Souza AI, Stewart S et al (2005a) Mössbauer study of glasses in meteorites: the D′Orbigny angrite and Cacharí eucrite. Hyperfine Interact 166(1–4): 543–547

Abdu Y, Souza AI, Stewart S, López A, Varela M, Kurat G, Scorzell R (2005b) Glasses in the d′orbigny and cacharí meteorites: a Mössbauer study. Meteorit Planet Sci 40:A13

Acevedo RD, Moreiras D, Ordaz J, Rodríguez-Terente LM (2005) Microanálisis químicos cuantitativos en la CO L_6 La criolla (república argentina). In: Martínez-Frías J, Madero Jarabo J (eds) Meteoritos y geología planetaria, pp. 93–103. Museo de las Ciencias de Castilla-La Mancha. Madrid, España

Acevedo RD, Rocca MCL (2005) Revisión crítica de los probables cráteres de impacto situados en territorio argentino. Actas del XVI Congreso Geológico Argentino III, La Plata, pp. 627–634

Acevedo RD, Rocca M (2008) Revisión y catálogo actualizado de los meteoritos hallados en territorio argentino. Actas del XVII Congreso Geológico Argentino III, San Salvador de Jujuy, pp. 1317–1318

Acevedo RD, Rocca M (2010) Los meteoritos argentinos son bienes culturales de la Nación. V Taller de Ciencias Planetarias, Sesión 7, Pósters II, La Plata, p. 26

Acevedo RD, Rocca M (2011) Catálogo de los meteoritos hallados en territorio argentino. Revista Historia Nat, Tercera Serie 1:17–34

Acevedo RD, Subías I (2012) A new old meteorite find from Claromecó, Argentina. 75th annual meeting of the meteoritical society. Meteorit Planet Sci 47, Supplement, s1, 5040. Cairns, Australia

Acevedo RD, Valín-Alberdi ML, Villar LM (2002) Hallazgo del mineral fosfuro de Níquel en una octahedrita IAB de Rubín de Celis (Campo del Cielo, Argentina). Resúmenes Primer Congreso Ibérico de Meteoritos y Geología Planetaria, Museo de las Ciencias de Castilla-La Mancha (Cuenca, España)

Ameghino F (1914) Aerolito fósil. Obras completas y Correspondencia Científica 2:276–279

Benet JL (1961) El meteorito de Arroyo Aguiar. Universidad Nacional del Litoral, Primera Reunión Trabajos Comunicaciones en Ciencias Naturales. Geografía, pp 323–326

Cabanillas ED, Palacios TA (2006) An hexahedrite meteorite from the campo del cielo fall. Planet Space Sci 54(3):303–309
Carnevali F (1953) El meteorito de vera. Revista Minera 21(1):29–32
Cassidy WA (1967) Meteorite fields studies at campo del cielo. Sky Telescope 34(1):4–10
Cassidy WA (1968) Meteorite impact studies at campo del cielo. In: French BM, Short NM (eds) Shock metamorphism of natural materials mono books corporation, Baltimore, Argentine, pp 117–128
Cassidy WA (1971) A small meteorite crater = structural details. J Geophys Res 76:3896–3912
Cassidy WA, Renard ML (1996) Discovering research value in the campo del cielo, Argentina, Meteorite craters. Meteorit Planet Sci 31:433–448
Cassidy WA, Villar LM, Bunch TE, Kohman TP, Milton DJ (1965) Meteorite and craters of campo del cielo, Argentina. Science 149:1055–1064
Clarke RS Jr (1974) The meteoritical bulletin, 52. Meteoritics 9:101–121
Clarke RS Jr (1975) The meteoritical bulletin, 53. Meteoritics 10:133–158
Connolly Jr HC, Zipfel J, Grossman JN, Folco L, Smith C, Jones R, Righter K, Zolensky M, Russell SS, Benedix G, Yamaguchi A, Cohen BA (2006) The meteoritical bulletin, N° 90, 2006 March. 41. Meteorit Planet Sci N 9:1383–1418
Connolly Jr HC, Zipfel J, Folco L, Smith C, Jones R, Benedix G, Righter K, Yamaguchi A, Chennaoui AH, Grossman JN (2007) The meteoritical bulletin. Meteorit Planet Sci N 91 March 42, N 3, 413–466
Eugster O, Busemann H, Kurat G, Lorenzetti S, Varela ME (2002) Noble gases in and CRE of the d'orbigny angrite. Meteorit Planet Sci 37:A44
Fletcher L (1899) On a mass of meteoric iron from the neighbourhood of caperr. Mineral Mag 12(56):167–170
Fossa Mancini E (1947a) Algunas observaciones sobre el meteorito de el perdido, (partido de coronel dorrego, provincia de buenos aires). Notas Museo de la Plata 12. Geología. N 45:109–142
Fossa Mancini E (1947b) La composición mineralógica de la piedra meteórica de indio rico. notas museo de la plata 12. Geología. N 46:143–156
Fossa Mancini E (1948) Hexaedritas halladas en la Argentina. Notas museo de la plata 13. Geología 50:107–112
Garvie LAJ (2012) The meteoritical bulletin, N 99, April 2012. MAPS 47(11):E1–E52
Giacomelli L.O. (1962) Meteoritos hallados en la patagonia. Argentina, Austral. Año XXXIV, N°14 370–371
Giacomelli LO (1969) Guía de meteoritos de la. Revista Museo Argentino de Ciencias Naturales, Geología, Argentina. Tomo 7, N 1
Gordillo CE (1959) El meteorito de distrito quebracho. Museo Entre Ríos Dirección de Prensa. Cultura N 1 (s/n)
Graham AL (1986) The meteoritical bulletin 64:310
Grossman JN (1996) Catalogs and inventories. The meteoritical bulletin, N 80*, 1996 July. Meteorit Planet Sci 31:A175–A180
Grossman JN (1997) Catalogs and inventories The meteoritical bulletin, N 81*, 1997 July. Meteorit Planet Sci 32:A159–A166
Grossman JN (1998) The meteoritical bulletin, N 82*, 1998 July. Meteorit Planet Sci 33:A221–A239
Grossman JN (1999) The meteoritical bulletin, N 83*, 1999 July. Meteorit Planet Sci 34:A169–A186
Grossman JN, Zipfel J (2001) The meteoritical bulletin, N 85*, 2001 September. Meteorit Planet Sci 36:A293–A322
Herrero DE (1908) El hierro meteórico de Puerta de Arauco. Revista Museo de La Plata 15(2): 84–90
Herrero DE (1911) Nota sobre el meteorito el perdido. Revista Museo de La Plata 18(2): 29–33
Herrero Ducloux E (1914) Nota sobre el meteorito carbonoso de nogoyá. Anales Museo Nacional de Historia Natural. Buenos Aires. 26. Mineralogía. Petrografía. N 3:99–116

Herrero Ducloux E (1925) Nota sobre el meteorito de la colina. Anales Museo Nacional de Historia Natural. Buenos Aires. 33:287–295
Herrero DE (1926) Nota sobre el meteorito de santa isabel. Revista Facultad de ciencias químicas La Plata 4, parte 1, 23–29
Herrero DE (1928a). Datos químicos sobre la piedra meteórica hinojo. Revista de la facultad química farmacia la plata 5, parte 2, 1–7
Herrero Ducloux E (1928b) Nota sobre el meteorito de pampa del infierno. Anales Museo de Historia Natural. Buenos Aires 34. Mineralalogía. Petrografía. N 8:597–601
Herrero Ducloux E (1929) Datos sobre la piedra meteórica de cacharí. Revista Facultad de Química y Farmacia, La Plata, 5. Parte 2:13–23
Herrero Ducloux E (1939) Nota sobre el meteorito de Águila Blanca. Notas Museo de La Plata 4. Geología. N9:353–360
Herrero Ducloux E (1940) Nota sobre el meteorito de gualeguaychú. Anales Museo Argentino de Ciencias NaturaleS Buenos Aires. 40. Mineralogía. Petrografía. N 14:123–127
Herrero Ducloux E (1942) Nota sobre el meteorito de san carlos notas museo de la plata 7, Geología. N 19:123–128
Herrero Ducloux E (1945a) Nota sobre el hierro meteórico de Ñorquín. Notas Museo de La Plata 10. Geología. N 40:163–164
Herrero Ducloux E (1945b) Nota sobre el meteorito de tostado. Anales Museo Argentino de Ciencias Naturales Buenos Aires 10. Geología. N 41:165–169
Herrero Ducloux E (1949) Notas sobre el hierro meteórico de campamento dadín. Notas Museo de La Plata 14. Geología N 54:177–179
Herrero DE, Pastore F (1929) El meteorito de renca. Revista Facultad Química Farmacia La Plata 5, parte 2, 111–120
Herrero Ducloux E, Pastore F (1930) El Meteorito de ishtilart. Anales Museo Argentino de Ciencias NaturaleS Buenos Aires 36. Mineralogía. Petrografía. N 9:313–330
Jagoutz E, Jotter R, Kubny A, Varela ME, Zartman R, Kurat G, Lugmair GW (2003) Cm?-U-Th-Pb isotopic evolution of the D'Orbigny angrite. Meteorit Planet Sci 38:A81
Jotter R, Jagoutz E, Varela ME, Zartman R, Kurat (2002) Pb isotopes in glass and carbonate of D'Orbigny angrite. Meteorit Planet Sci 37:A73
Jotter R, Jagoutz E, Varela ME, Zartman R, Kurat (2003) Lead isotopic study of glasses from the D'Orbginy angrite. Meteorit Planet Sci, 38:A53
Kubny A, Banerjee A, Jagoutz E, Varela ME, Brandstaetter F, Kurat G (2003) Some properties of an unusual glass and carbonate in the D'Orbigny angrite. Meteorit Planet Sci 38:A24
Kurat G, Ntaflos T, Brandstaetter F, Varela ME, Sylvester P, Nazarov M (2001) Trace element contents of major phases of the D'Orbigny angrite. Meteorit Planet Sci 36:A108
Kurat G, Varela ME, Brandstaetter F, Zinner E (2003) Large plates of anorthite-olivine growths in the D'Orbginy angrite. Meteorit Planet Sci, 38:A 57
Kurat G, Varela ME, Brandstaetter F, Weckwert G, Clayton R, Webwer L, Schultz L, Wäsch E, Nazarov M (2005) D'Orbigny: a non-igneous angritic achondrite. Geochim Cosmochim Acta 68:1901–1921
Levine J, Arazi A, Faestermann T, Fernández Hiello JO, Korschinek G, La GAMG, Negri A, Rugel G, Steier P, Wallner A (2007) Terrestrial age determination of an achondrite from Río Cuarto. Lunar Planet Sci, Argentina, 38, 1362.pdf
Lewis ChF, Moore CB, Hillar NA (1974) El sampal, a new meteorite from Argentina. Meteoritics 9:365–366
Liberman RG, Fernández NJO, Di TML, Fifield LK, Masarik J, Reedy RC (2002) Campo del cielo iron meteorites: sample shielding and meteoroid's preatmospheric size. Meteorit Planet Sci 37:295–300
Mittlefehldt DW, Killgore M, Lee MT (2002) Petrology and geochemistry of D'Orbigny, geochemistry of Sahara 99555, and the origin of angrites. Meteorit Planet Sci 32:345–369
Morello O, Anesa J (2000) Hallazgo de siliciuros de Fe en el meteorito El Aybal, Salta V Congreso de Mineralogía y Metalogenia, Actas, 495–496, La Plata, Argentina

Nágera JJ (1926). Los hoyos del campo del cielo y el meteorito. Dirección general de minas geología e hidrología, Ministerio de Agricultura de la Nación, Publicación 19, Buenos Aires

Nazarov MA, Brandstaetter F, Kurat G, Ntaflos T (1998) Chemistiy of P-rich sulfides in murchison, Cold bokkeveld and nogoya CM chondrites. Lunar Planet Sci, XXIX, abstract 1628

Olsacher J (1931) El meteorito de salavina. Revista Universidad Nacional de Córdoba, Año 18. N 9–10:430–446

Olsacher J (1939) El meteorito de hierro el simbolar. Boletín Facultad de Ciencias Exactas Físicas y Naturales de Córdoba. Año 2. N 3–4:79–88

Olsacher J (1951a) Condrita de achiras. Boletín Academia Nacional de Ciencias de Córdoba 39:261–267

Olsacher J (1951b) Condrita de quebrada de la aguada. Boletín Academia Nacional de Ciencias de Córdoba 39:268–273

Pastore F (1925a) Aerolito de la colina: anales museo nacional de historia natural. buenos aires 33. Mineralogía, Petrografía 6:297–306

Pastore F (1925b) Aerolito de santa isabel: anales museo nacional de historia natural. buenos aires 33. Mineralogía, Petrografía. 6:306–309

Renard ML, Cassidy WA (1971) Entry trajectory and orbital calculations for the crater 9 meteorite, campo del cielo. J Geophys Res 76:7916–7923

Rivas S, Rivas TEC, Ovando N (1976) Nuevo meteorito del lago nahuel huapi. Revista Minera 32(18):1975–1976

Rocca MCL (2005) La criolla meteorite shower, Entre Rios: Meteoroid's Heliocentric Orbit. 68th Annual Meeting of the Meteoritical Society, Gatlinburg, Tennessee, USA, Meteorit Planet Sci Supplement, Argentina, Abstract 5003, September 2005

Rocca MCL (2006) A catalogue of large meteorite specimens from campo del cielo. Meteorit Planet Sci, Chaco, Argentina, Supplement 41, August, A151

Rönick R, Ntaflos T, Palme H (2011) A posible paired fall of nicolás levalle from Argentina and its P in chondruleS 74th Annual Meeting of the Meteoritical Society. London, Meteorit Planet Sci 46(Supplement):5376

Russell S, Zipfel J, Folco, L, Jones R, Grady MT, Grossman JN (2003) The Meteoritical Bulletin. Meteorit Planet Sci, Supplement N 87, 2003 July 38(7), A189–A248

Russell S, Folco L, Grady M, Zolensky M, Jones R, Righter K, Zipfel J, Grossman JN (2004) The meteoritical bulletin. Meteorit Planet Sci N 88 (39), A215–A272

Salpeter, E.W, SJ (1957). The vatican collection of meteorites. Specola, Vaticana

Scientific event alert network bulletin (1985) 10(2):16

Scorzelli R, Souza AI, Stewart S, Varela ME, Kurat G (2004) Druses pyroxenes in D'Orbigny: a Mössbauer spectroscopy study. Meteorit Planet Sci, Supplement, A96

de Serralonga AME (1969) Descripción de un meteorito caído en el departamento ameghino, provincia de chubut. Actas Jornadas Geológicas Argentinas I:519–524

Stelzner T, Heide K, Bischoff A, Weber D, Merchel S, Herpers U, Faestermann T, Knie K, Korschinek G, Kubik PW, Suter M, Neumann S, Michel R, Scherer P, Schultz L, Jull AJT (1997) Rincon: a new L_6 chondrite find from Argentina, Chem. Erde 57:297

Teruggi ME (1968) El meteorito condrítico Chajarí. Revista del Museo de La Plata, Geología, VI 21 p

The American Mineralogist (1955) 7 El Simbolar. Argentina. 40(9–10):937

The Permanent Commission on Meteorites of the International Geological Congress (1958a) Meteorites not included in the prior-Hey catalogue of meteorites (1953) The meteoritical bulletin, 8:1–10. Moscow, URSS

The Permanent Commission on Meteorites of the International Geological Congress (1958b) Meteorites not included in the prior-Hey catalogue of meteorites (1953) The meteoritical bulletin, 10:1–3. Moscow, URSS

The Permanent Commission on Meteorites of the International Geological Congress (1961) Meteorites not included in the prior-Hey catalogue of meteorites (1953) The meteoritical bulletin, 21:1–3. Moscow, URSS

The Permanent Commission on Meteorites of the International Geological Congress (1962a) Meteorites not included in the prior-Hey catalogue of meteorites (1953) The Meteoritical Bulletin, August 1962, 24:1–6. Moscow, URSS

The Permanent Commission on Meteorites of the International Geological Congress (1962b) Meteorites not included in the prior-Hey catalogue of meteorites (1953) The meteoritical bulletin, 25:1–3. Moscow, URSS

The Permanent Commission on Meteorites of the International Geological Congress (1964a) Meteorites not included in the prior-Hey catalogue of meteorites (1953) The Meteoritical Bulletin, 29:1–4. Moscow, URSS

The Permanent Commission on Meteorites of the International Geological Congress (1964b) Discovery of Arbol Solo stony meteorite, Argentina. The meteoritical bulletin, 32:1–6, Moscow, URSS

The Permanent Commission on Meteorites of the International Geological Congress (1965) The meteoritical bulletin, 33:1–6. Moscow, URSS

Toselli AJ, Brodtkorb M (1973) El meteorito Casilda. Acta Geológica Lilloana 12:135–153

Turone OA (2001) The Balcarce meteorite. Meteorite 7(3):40

Turone OA (2002) The Caperr meteorite. Meteorite 8(2):34–36

Varela ME, Kurat G, Ntaflos T, Brandstaetter F, Sylvester P (2001) Trace elements in glass of the D'Orbigny angrite. Meteorit Planet Sci 36:A211

Varela ME, Kurat G, Zinner E, Brandstaetter F (2003a) Glasses in the D'Orbigny angrite. Meteorit Planet Sci 38:A59

Varela ME, Kurat G, Zinner E, Métrich N, Brandstaetter F, Ntaflos T, Sylvester P (2003b) Glasses in D'Orbigny angrite. Geochim Cosmochim Acta 67:5027–5046

Varela ME, Kurat G, Zinner E, Hoppe P, Ntaflos T, Nazarov M (2005) The non-igneous genesis of angrites: support from trace element distribution between phases in ĎOrbigny. Meteorit Planet Sci 40(3):409–430

Varela ME, Magnelli D, Morello O (2010) Berduc y "Santa Lucía": condritas ordinarias finds en V Taller de Ciencias Planetarias. Resúmenes, Argentina. Sesión 11, Meteoritos y asteroides, La Plata: pp 18–19

Villar LM (1968) La dispersión meteorítica en la y Chile. Ciencia e Investigación (julio), Argentina: 302–314

Weisberg MK, Smith C, Benedix G, Herd CDK, Righter K, Haack H, Yamaguchi A, Chennaoui Aoudjehane H, Grossman JF (2009) The meteoritical bulletin. Meteorit Planet Sc, N 96, September 2009i, 44(9):1355–1397

Weisberg MK, Smith C, Benedix G, Herd CDK, Righter K, Haack H, Yamaguchi A, Chennaoui AH, Grossman JF (2010) The meteoritical bulletin. Meteorit Planet Sci, N 97 45(3): 449–493

Wlotzka F (1992) The meteoritical bulletin, N 72*. Meteoritics 27(109–117):477–483

Wlotzka F (1994) The meteoritical bulletin, N 77*, 1994 November. Meteoritics 29:891–897

Wlotzka F (1995) The meteoritical bulletin, N 78*, 1995 November. Meteoritics 30:792–796

Wright SP, Vesconi MA, Spagnuolo MG, Cerutti C, Jacob RW. y Cassidy WA (2007) Explosion craters and penetration funnels in the Campo del Cielo, Argentina crater field. Lunar Planet Sci XXXVIII. 2017.pdf

Chapter 3
Bolivia

At present only three meteorites were found in this country where one of them is not geographically located (Fig. 3.1). Another possible meteorite (Bolivia iron, 21.25 kg., Smithsonian Institution collections) is surely paired with Pooposo octahedrite. The list is very brief. However this territory has many high altitude deserts that are optimal to recover old meteorites. No doubt in the near future many new specimens will be discovered in those wilderness areas.

Cochabamba. Coordinates not recorded. Cochabamba. Stone, carbonaceous chondrite, (CM_2), S_1. Find, date not recorded. Weight: 85 g. Kurat and Kracher (1975), Müller et al. (1977), Barber (1981), Nazarov et al. (2000).

Pooposo. 18° 20′ S, 66° 50′ W. Oruro. Iron, coarse octahedrite, (IAB complex). Find, 1910, 12 kg. Natural History Museum. Kichinka (1999).

Sevaruyo. 19° 22′ 04″ S, 66° 58′ 04″ W. Oruro. Stone. ordinary chondrite, (H_5), S_2 W_4. Olivine $Fa_{18.3}$, $Fs_{16.4}$. Find, 2001, 12.4 g. Bolivian National Museum, La Paz (Bolivia), Planetary Studies Foundation, Harper College, Palatine, Illinois (USA), NASA/George C. Marshall Space Flight Center, Alabama (USA). Kichinka (2002), Russell et al. (2002). Fig. 3.2.

R. D. Acevedo et al., *Catalogue of Meteorites from South America*,
SpringerBriefs in Earth System Sciences, DOI: 10.1007/978-3-319-01925-3_3,

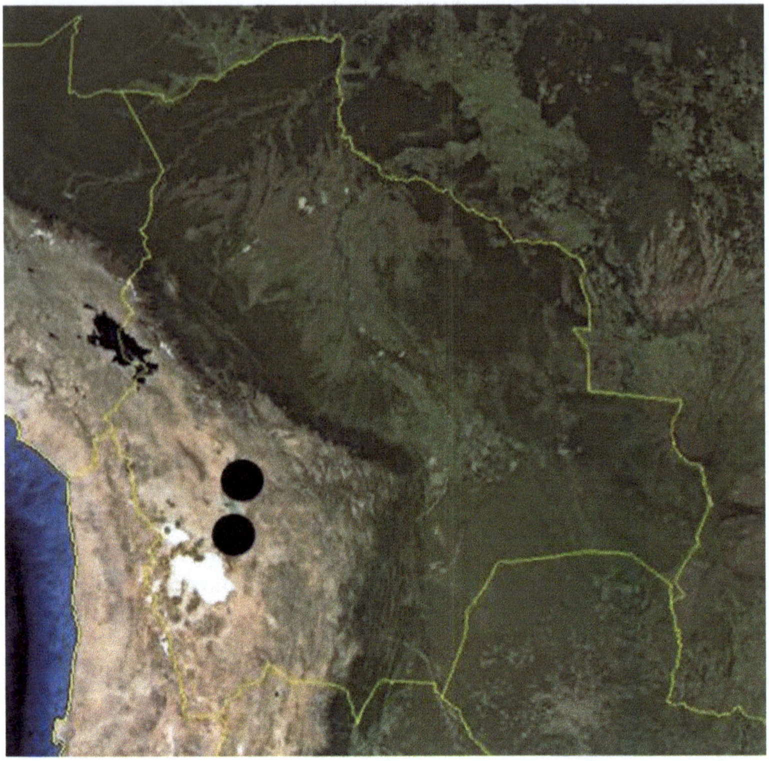

Fig. 3.1 Bolivian meteorites. Modified from © 2013 Inav/Geosistemas SRL

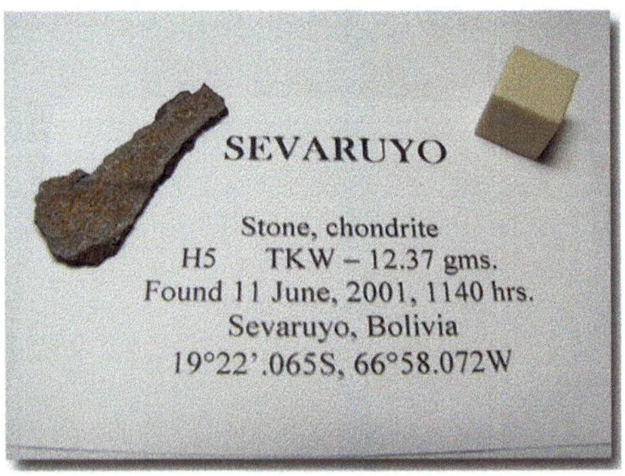

Fig. 3.2 Sevaruyo. Used by permission of Kevin Kichinka

Acknowledgments This brief overview of South American meteorites was partially funded by the Centro Austral de Investigaciones Científicas (CADIC) and the Consejo Nacional de Investigaciones Científicas y Tecnológicas (CONICET) of Argentina, the National Geographic/ Waitt, and The Planetary Society. We are also grateful to Jorge Rabassa for his remarks on the manuscript and to Ricardo Alonso for his valuable comments about some new and/or little-known meteorites.

References

Kurat G, Kracher A (1975) Preliminary report on the Cochabamba carbonaceous chondrite (abstract). Meteoritics 10:432–433

Müller WF, Kurat G, Kracher A (1977) Crystal Structure and Composition of Cronstedtite from the Cochabamba Carbonaceous Chondrite. Meteoritics 12:322

Barber DJ (1981) Matrix phyllosilicates and associated minerals in C2M carbonaceous chondrites. Geochim Cosmochim Acta 45:945–970

Nazarov MA, Kurat G, Brandstaetter F, Ntaflos T (2000) A Calcium Enrichment in Phosphorian Sulfides and Barringerites from the Cochabamba (CM) Chondrite. Meteoritics & Planetary Science, vol. 35(suppl):A117

Kichinka K (1999) No Stone Left Unturned in a Search for Bolivia's First Authenticated Meteorite. Meteorite 5(4):12–15

Kichinka K (2002) A Meeting of Heaven and Earth – Bolivia's First Authenticated Meteorite. Meteorite 8(2):14–19

Russell SS, Zipfel J, Grossman JN, Grady MM (2002) The Meteoritical Bulletin, No 86. Meteorit Planet Sci 37(supp):A157–A184

Chapter 4
Brazil

The first meteorites discovered in Brazil have unusual stories. There are reports that the Macau meteorite shower (seen to fall on 1836) caused massive damages (many oxen and cattle killed and houses destroyed) but only a small pieces of an ordinary chondrite were recuperated. Another unique meteorite, the achondrite Angra dos Reis, fell in 1869, very early in the morning, immersing into the bay of Angra dos Reis at a water depth of 2 m, in front of the church of Bom Fim (Good End). It is the first specimen ever found of the very rare basaltic meteorite type called "angrites" which are supposed to come from the surface of planet Venus.

So far, the astronomer E. M. Zucolotto has published some catalogs of Brazilian meteorites: Zucolotto (1999, 2004), Zucolotto et al. (1999), and Zucolotto and Antonello (2004a). Data for all specimens included in addition to known cases are described below (Fig. 4.1).

Angra dos Reis. 22° 58′ S, 44° 19′ W. Rio de Janeiro. Achondrite, angrite. Seen falling on January 20, 1869, was recovered a weight of 1.5 kg. Museu Nacional UFRJ, Rio de Janeiro (Brazil), American Museum of Natural History, New York (USA), Smithsonian Institution collections, Natural History Museum. Ma et al. (1976), Bell et al. (1977), Prinz et al. (1977), Wasserburg et al. (1977), Treiman (1988), Treiman et al. (1988), Crozaz and McKay (1990), Mittlefehldt et al. (1990), Lofgren and Lanier (1992), Nyquist et al. (1994), Yanai (1994), Riches et al. (2012). Fig. 4.2.

Avanhandava. 21° 27′ 37″ S, 49° 57′ 03″ W. São Paulo. Stone, ordinary chondrite, (H_4). Fall in 1952, 9.33 kg. Smithsonian Institution collections. Paar et al. (1976), Lux et al. (1981), Kessel et al. (2004), Kohout and Pesonen (2005), Kohout et al. (2006, 2007). Fig. 4.3.

Balsas. 07° 31′ 88″ S, 46° 02′ 47″ W. Maranhão. Iron, octahedrite (IIIAB). Find, 1974, 41 kg. Associação Carazinhense de Astronomia e Estudos Espacials (Rio Grande do Sul), Museu Nacional UFRJ, IGPP-UCLA. Grossman (1998), Zucolotto (1999). Fig. 4.4.

R. D. Acevedo et al., *Catalogue of Meteorites from South America*, SpringerBriefs in Earth System Sciences, DOI: 10.1007/978-3-319-01925-3_4,

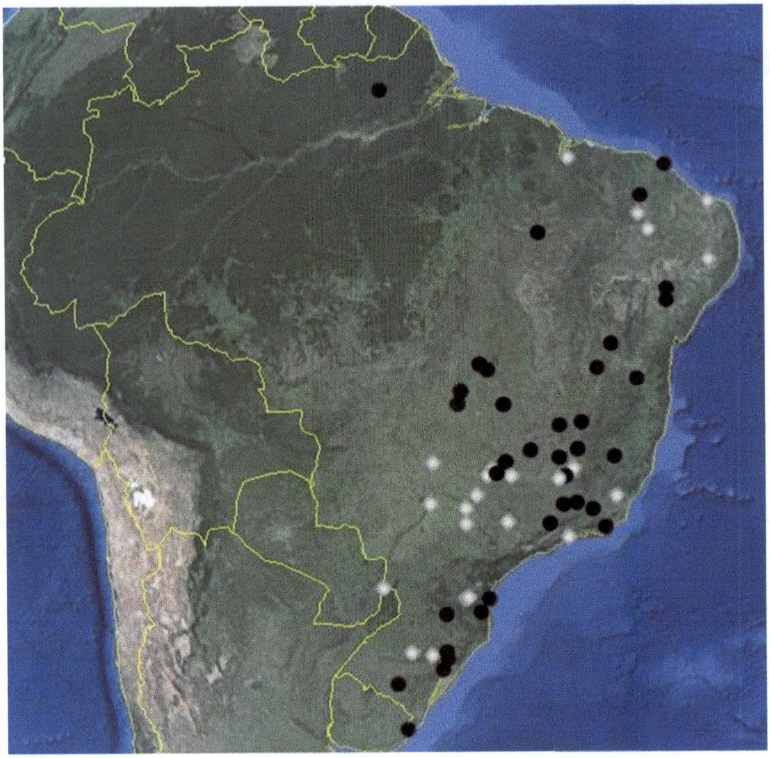

Fig. 4.1 Brazilian meteorites. Modified from © 2013 Inav/Geosistemas SRL

Fig. 4.2 Angra dos Reis. *Credit*: André Moutinho www.meteorito.com.br

Fig. 4.3 Avanhandava. *Credit*: Instituto Geologico, Secretaria de Estado do Meio Ambiente

Fig. 4.4 Balsas. *Credit*: Elizabeth Zucolloto

Fig. 4.5 Barbacena. *Credit*: André Moutinho

Barbacena. 21° 13′ S, 43° 56′ W. Minas Gerais. Iron, ungrouped. Ni = 10.5 wt %. Find, 1918, 9.03 kg. Smithsonian Institution collections, Natural History Museum. Curvello and Ferreira (1951), Zucolotto and Pinto (2000). Fig. 4.5.

Bendegó. 10° 07′ S, 39° 12′ W. Bahía. Iron, (IC). Found in 1784, 5.36 MT. Museu Nacional UFRJ, Smithsonian Institution collections, Natural History Museum. Sears (1963), Zucolotto et al. (1999), Rios and Carvalho (2009), Carvalho and Rios (2011), Belmonte et al. (2012). Fig. 4.6.

Blumenau. 26° 55′ 26″ S, 49° 03′ 32″ W. Santa Catarina. Iron, fine octahedrite, (IVA). Found in 1986. Unknown mass. Museu Nacional UFRJ, IGPP-UCLA. Grossman (1998). Fig. 4.7.

Bocaiúva. 17° 07′ S, 43° 49′ W. Minas Gerais. Iron, anomalous with silicate inclusions. Ni = 8.49 wt %. Find, 1965, 64 kg. Natural History Museum. Araujo

Fig. 4.6 Bendegó. *Credit*: André Moutinho www.meteorito.com.br

Fig. 4.7 Blumenau. *Credit*: André Moutinho www.meteorito.com.br

et al. (1983), Curvello et al. (1983), Graham (1984), Desnoyers et al. (1985), Malvin et al. (1985), Scorzelli, and Danon (1986), Funaki et al. (1987), Johnson et al. (1989), Liu et al. (2001). Fig. 4.8.

Campinorte. 14° 15′ 48″ S, 49° 09′ 55″ W. Goiás. Iron, coarse octahedrite? Found in 1992. A large pear shaped mass of about 2 tons with no fusion crust or regmaglypts. Specimens are deposited in the Museu Nacional UFRJ, IGPP-UCLA, University of California, Los Angeles (USA), and University of Alberta (Canada). Garvie (2012). Fig. 4.9.

Campos Sales. 07° 02′ S, 40° 10′ W. Ceará. Stone, ordinary chondrite, (L_5). Olivine Fa_{25}, pyroxene $Fs_{21.4}$ Meteorite shower fell on January 31, 1991, and 23.68 kg (~300 fragments) were collected. Centro Brasileiro de Pesquisas Físicas (Brazil), Natural History Museum. Wlotzka (1995). Fig. 4.10.

Fig. 4.8 Bocaiúva. *Credit*: Filipe Chaves, Museu de Mineralogia Prof. Djalma Guimaraes

Fig. 4.9 Campinorte. *Credit*: André Moutinho www.meteorito.com.br

Fig. 4.10 Campos Sales. *Credit*: Svend Buhl

Casimiro de Abreu. 22° 28′ S, 42° 13′ W. Rio de Janeiro. Iron, (IIIAB). Found in a farm in 1947 (the stone would have been found much earlier than it is registered and was kept underneath the owner bed). A 25 kg specimen, it is deposited in the Museu Nacional UFRJ, and Smithsonian Institution collections. Curvello (1950a). Fig. 4.11.

Conquista. 19° 51′ S, 47° 33′ W. Minas Gerais. Stone, olivine-broncite chondrite, (H_4). Olivine $Fa_{17.2}$, pyroxene $Fs_{15.2}$. 25.83 % total iron. Seen falling on

Fig. 4.11 Casimiro de Abreu. *Credit*: Elizabeth Zucolloto

Fig. 4.12 Conquista. *Credit*: André Moutinho www.meteorito.com.br

Fig. 4.13 Governador Valadares. *Credit*: The Macovich Collection Darryl Pitt

December, 1965. Total weight unknown, single specimen: 20.35 kg. Smithsonian Institution collections. Keil et al. (1978a, b), Graham et al. (1981). Fig. 4.12.

Cratheús (1931). 05° 15′ S, 40° 30′ W. Ceará. Iron. Octahedrite, (IVA). Found in 1914, 27.5 kg. Smithsonian Institution collection, Natural History Museum.

Cratheús (1950). Iron, plessitic octahedrite, (IIC). Found in 1909, 367 g. Smithsonian Institution collections, Natural History Museum. Curvello (1950b).

Governador Valadares. 18° 51′ S, 41° 57′ W. Minas Gerais. Achondrite, Martian nakhlite. Find, 1958, 158 g. Smithsonian Institution collection, Natural History Museum. Burragato et al. (1975), Berkley et al. (1979, 1980), Harvey and McSween (1991), Mittlefehldt and Lindstrom (1996), Mikouchi et al. (2004), Miyamoto (2004), Chevrier et al. (2011), Korochantseva et al. (2011). Fig. 4.13.

Fig. 4.14 Ibitira. *Credit*: Mike Bandli

Fig. 4.15 Indianópolis. *Credit*: André Moutinho www.meteorito.com.br

Ibitira. 20° S, 45° W. Minas Gerais. Achondrite, eucrite (monomict breccia). Fall in 1957, 2.5 kg. Smithsonian Institution collections, The Natural History Museum. Menezes (1957), Wilkening y Anders (1974), Steele and Smith (1976), Davis et al. (2001), Mittlefehldt (2005), Burbine et al. (2006), Bunch et al. (2009). Fig. 4.14.

Iguaraçu. 21° 12′ S, 51° 50′ W. Paraná. Stone. ordinary chondrite, (H_5). Fall, 1977, 1,200 g. Fundação Universidade de Maringá, Massey University, Palmerston North (New Zealand), British Museum, Natural History Museum. Brooks et al. (1990), Graham (1990).

Indianópolis. 19° 10′ S, 47° 50′ W. Minas Gerais. Iron. Octahedrite, (IIAB). It is similar in composition to the Santa Luzia meteorite. Rich in schreibersite. Found in 1989, 14.85 kg. Universidade de São Paulo (Brazil), Max-Planck Institut für Chemie, Mainz (Alemania). , Grossman (2000). Fig. 4.15.

Ipiranga (Lajeado). 25° 30′ S, 54° 30′ W. Paraná. Stone, olivine-bronzite chondrite, (H_6). Olivine Fo_{79}, bronzite En_{82}. Meteorite shower seen falling on December 27, 1972, weighing 7 kg altogether (an individual of 2.65 kg and a number of small stones). Smithsonian Institution collections. Clarke Jr. (1974), Cavarretta et al. (1975), Gomes et al. (1978a, b). Fig. 4.16.

Ipitinga. 00° 21′ N, 53° 49′ W. Pará. Stone, ordinary chondrite, (H_5). Olivine $Fa_{18.3}$. Found in 1989, 7 kg. Museu de Mineralogía, Belem, Pará (Brazil. Dreher et al. (1995), Wlotzka (1992). Fig. 4.17.

Itapicuru-Mirim. 03° 24′ S, 44° 20′ W. Maranhao. Stone. ordinary chondrite, (H_5). It was seen falling sometime in March, 1879, 2.02 kg. Smithsonian Institution collections, The Natural History Museum. Gomes et al. (1977b). Fig. 4.18.

Fig. 4.16 Ipiranga. *Credit*: Unknown

Fig. 4.17 Ipitinga. *Credit*: Unknown

Fig. 4.18 Itapicuru-Mirim. *Credit*: André Moutinho www.meteorito.com.br

Fig. 4.19 Itapuranga. *Credit*: Museu de Geociencias-IGc/USP

Itapuranga. 15° 35′ S, 50° 09′ W. Goiás Iron, coarse octahedrite, (IAB complex). Ni = 6.7 wt %. Year found unknown, 628 kg. Graham (1980), Kracher et al. (1980), Svisero et al. (1980). Fig. 4.19.

Itutinga. 21° 20′ S, 44° 40′ W. Minas Gerais. Iron, (IIIAB). Find, 1960, 3.2 kg. and a 110.76 g fragment named "Itumirim" by Nunes et al. (2010a, b). Smithsonian Institution collections. Fig. 4.20.

Lavras do Sul. 30° 48′ S, 53° 54′ W. Rio Grande do Sul. Stone, ordinary chondrite, (L_5), $S_{3/4}$ W_1. Olivine Fa_{25}, low-Ca pyroxene $Fs_{22.6}$. It is similar in composition to the Putinga meteorite. Found in 1985, ~1 kg. Museu Nacional do Rio de Janeiro, Centro Brasileiro de Pesquisas Físicas. Zucolotto et al. (2010a, b, 2012a), Garvie (2012).

Macau. 05° 12′ S, 36° 40′ W. Rio Grande do Norte. Stone, ordinary chondrite, (H_5). This meteorite shower fell on the 11th November, 1836, 1.5 kg. Smithsonian Institution collections, Natural History Museum. Gomes et al. (1977c). Fig. 4.21.

Mafra. 26° 10′ S, 49° 56′ W. Santa Catarina. Stone. ordinary chondrite, H_4 (Levi-Donati et al. 1976), $L_{3/4}$ (Lange et al. 1979). Fall, 1941, 600g. Smithsonian Institution collections. Sighinolfi and Gorgoni (1983).

Fig. 4.20 Itutinga. *Credit*: Museu de Ciencia e Tecnica da Escola de Minas/UFOP

Fig. 4.21 Macau. *Credit*: André Moutinho www.meteorito.com.br

Fig. 4.22 María da Fé. *Credit*: André Moutinho www.meteorito.com.br

María da Fé. 22° 18′ S, 45° 22′ W. Minas Gerais. Iron, fine octahedrite, (IVA). Bulk metal Ni = 7.45 %. Found, 1987, 18 kg. Museu Nacional UFRJ, IGPP-UCLA. Grossman (1999). Fig. 4.22.

Marília. 22° 15′ S, 49° 56′ W. São Paulo. Stone, olivine-bronzite chondrite, (H_4). Date of fall: October 5, 1971, 2.5 kg (7 specimens) were identified.

Fig. 4.23 Marília. *Credit*: André Moutinho www.meteorito.com.br

Fig. 4.24 Minas Gerais. *Credit*: André Moutinho www.meteorito.com.br

Fig. 4.25 Nova Petrópolis. *Credit*: André Moutinho www.meteorito.com.br

Smithsonian Institution collections, Natural History Museum. Avanzo et al. (1973), Shima et al. (1974), Rambaldi et al. (1979). Fig. 4.23.

Minas Gerais (a). 18° 30′ S, 44° W. Minas Gerais. Stone, ordinary chondrite, (L_6). Found in 1888, 1,224 g. Smithsonian Institution collections, Natural History Museum. Fig. 4.24.

Minas Gerais (b). Unknown coordinates in Minas Gerais. Stone, ordinary chondrite, (H_4), S_3 $W_{1/2}$. Olivine $Fa_{19.2\ \pm\ 2.0}$, pyroxene $Fs_{15.1\ \pm\ 5.6}$ $Wo_{1.4\ \pm\ 1.1}$ (Classifier: R. Bartoschewitz and P. Appe). Found in 2001, 42.6 g. Connolly et al. (2006).

Morro do Rócio. 27° S, 51° W. Santa Catarina. Stone, ordinary chondrite, (H_5). Found in 1928, 369 g. Natural History Museum. Wlotzka and Fredriksson (1980), Fredriksson and Wlotzka (1985), Wlotzka (1985).

Nova Petrópolis. 29° 26′ S, 50° 55′ W. Rio Grande do Sul. Iron, octahedrite, (IIIAB). Ni = 7.8 %. Found, 1967, 305 kg. Smithsonian Institution collections, Natural History Museum. Grunewaldt (1983), Graham (1984), Souza Azevedo et al. (1987), Funaki and Danon (1997). Fig. 4.25.

Fig. 4.26 Pará de Minas. *Credit*: Anne Black

Fig. 4.27 Parambu. *Credit*: Mateo Chinellato

Fig. 4.28 Paranaíba. *Credit*: Museu de Geociencias-IGc/USP

Palmas de Monte Alto. 14° 22′ 03″ S, 43° 1′ 22″ W. Bahía. Iron, octahedrite, (IIIAB). *Cf* Bendegó meteorite. Ni = 9.4 wt %. It was found before 1954, 97 kg. Museu Nacional UFRJ. Weisberg et al. (2009a, b), Zucolotto and Riff (2009).

Pará de Minas. 19° 52′ S, 44° 37′ W. Minas Gerais. Iron, (IVA). Collected in 1934, 116.3 kg. Smithsonian Institution collections. Curvello and Ferreira (1952), Fireman and Schwarzer (1957), Romig and Goldstein (1981). Fig. 4.26.

Paracutu. 03° 24′ S 39° 02′ W. Iron, (IAB complex). Found in 1980. Mass unknown.

Parambu. 06° 14′ S, 40° 42′ W. Ceará. Stone, ordinary chondrite, (LL_5). Fall in 1967, 2 kg. Barreto et al. (1973), Shima et al. (1973), Clarke (1974), Shima et al. (1974). Fig. 4.27.

Paranaíba (H_6 Cacilândia or Cassilândia). 19° 08′ S, 51° 40′ W. Mato Grosso do Sul. Stone, ordinary chondrite, (L_6). Seen falling on 1956 (Migomaspa fireball),

Fig. 4.29 Patos de Minas. *Credit*: Svend Buhl

Fig. 4.30 Patrimônio. *Credit*: Museu de Geociencias-IGc/USP

100 kg. Smithsonian Institution collections, Max Planck, Mainz (Germany). Keil et al. (1977), Zucolotto and Carvalho (2009). Fig. 4.28.

Patos de Minas 1 (Corrego Areado). 18° 35′ S, 46° 32′ W. Minas Gerais. Iron, hexahedrite, (IIAB). Found in 1925, 32 kg were recovered. Smithsonian Institution collections.

Patos de Minas 2 (Santa Fé). 18° 35′ S, 46° 32′ W. Minas Gerais. Iron, octahedrite (IAB complex). The highly unusual meteorite shows schreibersite and troilite inclusions in a chaotic disturbed lattice. Found in 2002, 198kg. Museu Nacional UFRJ, IGPP-UCLA (Los Ángeles). Scorzelli et al. (2003), Connolly et al. (2008), Pucheta and Cassino (2008). Fig. 4.29.

Patrimônio. 19° 32′ S, 48° 34′ W. Minas Gerais. Stone, ordinary chondrite, (L_6). Fall, 1950, 2.12 kg. Smithsonian Institution collections, Natural History Museum. Frederick (1956). Fig. 4.30.

Piedade do Bagre. 18° 56′ 30″ S, 44° 59′ W. Minas Gerais. Iron, medium octahedrite. Find, 1922, 59 kg. Smithsonian Institution collections, Natural History Museum. Spencer and Hey (1930). Fig. 4.31.

Pirapora (Angra Dos Reis hexaedrite). 17° 18′ S, 45° W. Minas Gerais. Iron, (IIAB) showing a typical hexahedrite structure with Neumann bands. Find, 1888,

Fig. 4.31 Piedade do Bagre. *Credit*: Steve Arnold

Fig. 4.32 Putinga fall. Drawing G. Schinke

Fig. 4.33 Quijingue. *Credit*: Mateo Chinellato

6.18 kg. Smithsonian Institution collections, Natural History Museum. Curvello (1954, 1958), Axon and Waine (1972), Scorzelli et al. (2003).

Porto Alegre. 30° 01′ 59″ S, 51° 13′ 48″ W. Rio Grande do Sul. Iron, (IIIE). Find, 2005, 200 kg. Museu Nacional do Rio de Janeiro, IGPP-UCLA. Garvie (2012).

Putinga. 29° 02′ S, 53° 03′ W. Rio Grande do Sul. Stone, ordinary chondrite, (L_6). Fall, 1937, 300 kg. Smithsonian Institution collections, Natural History Museum. Symes and Hutchinson (1970), Keil, et al. (1978), de Oliveira et al. (1988), Zucolotto (1999), Kuhn (2008), Antonello et al. (2010a). Fig. 4.32.

Quijingue. 10° 45′ S, 39° 13′ W. Bahía. Stony-Iron, pallasite. Found, 1984. 59 kg. Museu Nacional do Rio de Janeiro, IGPP-UCLA, Smithsonian Institution collections. Grossman (1999), Vianna Cautinho (1999). Fig. 4.33.

Rio do Pires. 13° 07′ 24″ S, 42° 17′ 19″ W. Bahía. Stone, ordinary chondrite, (L_6). Collected in 1991, 118 g. Wlotzka (1994), Zucolotto and Antonello (2001). Fig. 4.34.

Rio Negro. 26° 06′ S, 49° 48′ W. Bahía. Stone, ordinary chondrite, L_4. Fall, 1934. 1,310 g. Smithsonian Institution collection, Natural History Museum. Fodor et al. (1977). Fig. 4.35.

Fig. 4.34 Rio do Pires. *Credit*: André Mouthino www.meteorito.com.br

Fig. 4.35 Rio Negro. *Credit*: Unknown

Fig. 4.36 Sanclerlândia. *Credit*: Museum of the Institute of Geosciences, University of Brasília

Sanclerlândia. 16° 13′ S, 50° 18′ W. Goiás. Iron, (IIIAB). Find, 1971, 276 kg. Graham (1980). Fig. 4.36.

Santa Bárbara. 29° 12′ S, 51° 52′ W. Rio Grande do Sul. Stone, ordinary chondrite, (L_4). Fall, 1873, 400 g. Natural History Museum. Berkley et al. (1978). Fig. 4.37.

Fig. 4.37 Santa Bárbara. *Credit*: Fernlea meteorites

Fig. 4.38 Santa Catharina. *Credit*: Stephan Decker

Fig. 4.39 Santa Luzia. *Credit*: Museu de Geociencias-IGc/USP

Santa Catharina. 26° 13′ S, 48° 36′ W. Santa Catarina. Iron, (IAB ungrouped). Find, 1875, 7 MT. Smithsonian Institution collections, Natural History Museum. Bowles et al. (1978), Danon et al. (1979), Jago (1979), Miller and Russell (1992), Van Tassel et al. (1992), Goldstein et al. (1998), Brandstätter et al. (2003). Fig. 4.38.

Santa Luzia. 16° 16′ S, 47° 57′ W. Goiás. Iron, (IIAB). Find, 1921, 1.92 MT. Smithsonian Institution collections, Natural History Museum. Meen (1939), Curbello (1950c), Clarke Jr. et al. (1984). Fig. 4.39.

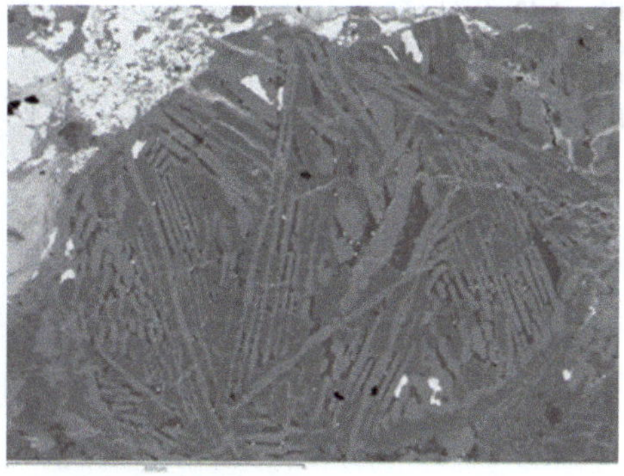

Fig. 4.40 Santa Vitória do Palmar. SEM image. *Credit*: María Luz Valín

Fig. 4.41 São João Nepomuceno. *Credit*: André Moutinho www.meteorito.com.br

Santa Vitória do Palmar. 33° 30′ 34″ S 53° 24′ 39″ W. Rio Grande do Sul. Stone, ordinary chondrite, (L_3), S_4 W_2. Find, 2003, 50.4 kg. Museum für Naturkunde, Berlin (Germany). Buhl and Greshake (2006), Pinheiro et al. (2006), Connolly et al. (2007), Zucolotto and Antonello (2008). Zucolotto et al. (2010a, b). Fig. 4.40.

São João Nepomuceno. 21° 33′ S, 43° 01′ W. Minas Gerais. Iron, (IVA anomalous). Find, 1960, 15.3 kg. Smithsonian Institution collections, Natural History Museum. Graham (1980), Wang et al. (2001), Zucolotto and Antonello (2004), Zucolotto et al. (2011b), Zucolotto and Monteiro (2012). Fig. 4.41.

São José do Rio Preto. 20° 48′ 36″ S, 49° 22′ 50″ W. São Paulo. Stone, ordinary chondrite, (H_4). Fall, 1962, 927 g. Smithsonian Institution collection.

Saulo Gomes. 22° 11′ S, 47° 42′ W. Buritizal. São Paulo Stone, ordinary chondrite, ($L_3/_4$). Fell in August 1967. Three artifacts of mass unknown. Universidade Estadual Paulista "Julio de Mesquita Filho" (UNESP) (Brazil). Zanardo et al. (2011). (Not officially listed yet in the Meteoritical Bulletin Database). Fig. 4.42.

Serra de Magé. 08° 23′ S, 36° 46′ W. Pernambuco. Achondrite, eucrite. Seen falling on 1923, 1800 g were rescued. Smithsonian Institution collections, Natural History Museum. Carver and Anders (1969, 1970), Harlow et al. (1977, 1979), Treiman et al. (2004), Antonello et al. (2010b). Fig. 4.43.

Fig. 4.42 Saulo Gomes. *Credit*: Antenor Zanardo, Universidade Estadual Paulista

Fig. 4.43 Serra de Magé. *Credit*: André Moutinho www.meteorito.com.br

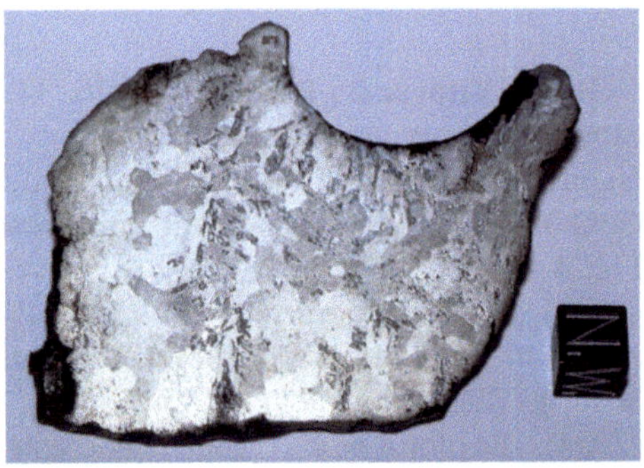

Fig. 4.44 Soledade. *Credit*: André Moutinho www.meteorito.com.br

Sete Lagoas. 19° 28′ S, 44° 13′ W. Minas Gerais. Stone, ordinary chondrite, (H_4). Fall, 1908, 350 g. Gomes and Keil (1977).

Soledade. 29° 03′ S, 51° W. Rio Grande do Sul, Brasil. Iron, octahedrite, (IAB complex). Ni = 6.7 wt %. Find, 1986, 68 kg. IGPP-UCLA, Smithsonian Institution collections. Graham (1988), Paduani et al. (2005). Fig. 4.44.

Uberaba. 19° 49′ S, 48° 47′ W. Minas Gerais. Stone, ordinary chondrite, (H_5). Fall, 1903, 40 kg. Smithsonian Institution collections, Natural History Museum. Gomes et al. (1977a). Fig. 4.45.

Uruaçu. 14° 32′ S, 48° 46′ W. Goiás. Iron, (IAB). Find, 1992, 72.5 kg. Museu Nacional do Rio de Janeiro, IGPP-UCLA. Russell et al. (2002). Fig. 4.46.

Fig. 4.45 Uberaba. *Credit*: Museu de Geociencias-IGc/ USP

Fig. 4.46 Uruaçu. *Credit*: Mike Farmer

Fig. 4.47 Varre-Sai. *Credit*: André Moutinho www.meteorito.com.br

Varre-Sai. 20° 51′ S, 41° 44′ 01″ W. Rio de Janeiro. Stone, ordinary chondrite, (L_5), S_4 W_0. Olivine $Fo_{0.75}$ and low-Ca pyroxene ($Fs_{21.66}$ $En_{76.48}$ $Wo_{1.49}$). A bright fireball was observed at noon, on June 19, 2010, N of Rio de Janeiro and S of Espirito Santo state. At least five masses (total ~2.5 kg) were recovered in both states. Museu Nacional do Rio de Janeiro and Centro Brasiliero de Pesquisas Fisicas, Rio de Janeiro (Brazil). Zucolotto et al. (2011a, 2012b), Garvie (2012). Fig. 4.47.

Fig. 4.48 Vitória da Conquista. *Credit*: André Moutinho www.meteorito.com.br

Veríssimo. 19° 44′ S, 48° 19′ W. Minas Gerais. Iron, (IIIAB). Collected in 1965, 14 kg.

Vitória da Conquista. 14° 50′ 19″ S, 40° 50′ 10″ W. Bahía. Iron, (IVA). The composition (by INAA) of the metal is Ni = 9.4 %, belonging to the IVA group. Find, 2007, 10.5 kg. Weisberg et al. (2009a, b), Zucolotto and Riff (2009). Fig. 4.48.

References

Antonello LL, Zucolotto ME, Scorzelli RB, Roisenberg A, Varela ME, Souza Azevedo I (2010a) Estudo comparativo entre os meteoritos Lavras do Sul e Putinga (Petrografia e Mineralogia). In: Congresso Brasileiro de Geologia, 45, 2010, Belém. Anais. Belém: Sociedade Brasileira de Geologia, 1 p. CD-ROM. (pap. 1557)

Antonello LL, Zucolotto ME, Scorzelli RB, Souza Azevedo I (2010b) Petrografia e Mineralogia do Meteorito Serra de Magé Revisado (8°23′S; 36°46′W). In: Congresso Brasileiro de Geologia, 45, Belém. Anais. Belém: Sociedade Brasileira de Geologia, 1 p. CD-ROM. (pap. 1583)

Araujo SI, Danon J, Scorzelli RB, Souza Azevedo I, Galvao da Silva E (1983) Mössbauer Study of Silicates and Metal Phases of the Bocaiuva Meteorite. Meteoritics 18:261

Avanzo PE, Levi-Donati GR, Sighinolfi GP (1973) The Marília Meteorite shower. A preliminary report. Meteoritics 8:141–147

Axon HJ, Waine CV (1972) A metallographic study of some hexaedrites. Mineral Mag 38:725–735

Barreto A, Fonseca de Mello Z, Levi-Donati GR, Sighinolfi GP (1973) The Meteorite Shower of Parambu, Ceará State, Brazil: Mineralogy and Petrology (Abstract). Meteoritics 8:324

Bell PM, Mao HK (1977) Crystal-Field Spectra of Fassaite from the Angra Dos Reis Meteorite. Abstracts Lunar Planet Sci Conf 8:85

Belmonte SLR, Zucolotto ME, Fontes RC, dos Santos JRL (2012) 3-D virtual and physical reconstruction of Bendegó Iron. 75th annual meeting of the meteoritical society, Cairns (Australia). Meteorit Planet Sci Supplement:5149

Berkley JL, Keil K, Gomes CB, Curvello WS (1978) Studies of Brazilian meteorites XII. Mineralogy and petrology of the Santa Bárbara, Rio Grande do Sul, chondrite. Anais da Academia Brasileira de Ciências 50(1):191–196

Berkley JL, Keil K, Prinz M, Gomes CG (1979) The Governador Valadares Nakhlite and its Relationship to Other Nakhlites. Lunar Planet Sci 10:101–103

Berkley JL, Keil K, Prinz M (1980) Petrology and origin of Governador Valadares and other nakhlites. In: Proceedings of the eleventh lunar and planetary science conference, Lunar and Planetary Institute, Houston, pp 1089–1102

Bowles JS, Hatherly M, Malin AS (1978) FeNi superlattice formation by corrosion of Santa Catharina meteorite. Nature 276:168–169

Brandstätter F, Nazarov MA, Kurat G (2003) Barringerite from the Santa Catharina ungrouped iron meteorite. Lunar Planet Sci 34:1681

Brooks RR, Hoashi M, Fernandes de Lima E, Dos Santos ML, Ryan DE, Holzbecher J, Johnston JH, Kawachi Y, Sipiera PP (1990) The Iguaracu H5 chondrite. A fall from Parana State, Brazil. Meteoritics 25:231–232

Buhl S, Greshake A (2006) Santa Vitoria do Palmar, 33,304 amigos and one meteorite: a new meteorite from Brazil. Meteorite 4:12–16

Bunch TE, Irving AJ, Rumble D, Korotev RL, Wittke JH, Sipiera PP (2009) Northwest Africa 2824: another Eucrite-like sample from the Ibitira Parent Body? 72nd annual meeting of the meteoritical society, Nancy, France. Meteorit Planet Sci Supplement:5367

Burbine TH, Dyar MD, Seaman SJ, McCoy TJ (2006) Water content of nominally anhydrous minerals in the Ibitira Eucrite. 37th annual lunar and planetary science conference, League City, Texas (USA), p 2220

Burragato F, Cavarretta G, Funiciello R (1975) The new Brazilian achondrite of Governador Valadares (Minas Gerais). Meteoritics 10:374

Carvalho WP, Rios DC (2011) Meteorite Bendegó, a Brazilian Comics Super Hero. 74th annual meeting of the meteoritical society, London. Meteorit Planet Sci Supplement:5111

Carver EA, Anders E (1969) Serra de Magé: an odd meteorite. Meteoritics 4:267–268

Carver EA, Anders E (1970) Serra de Magé: a meteorite with an unusual history. Earth Planet Sci Lett 8:214

Cavarretta G, Fornaseri M, Funiciello R, Tolomeo L (1975) The chondritic shower of Lajeado Ipiranga, Paraná, Brazil. Meteoritics 10:380

Chevrier V, Lorand JP, Sautter V (2011) Sulfide petrology of four nakhlites: Northwest Africa 817, Northwest Africa 998, Nakhla, and Governador Valadares. Meteorit Planet Sci 46(6):769–784

Clarke RS Jr (1974) The Meteoritical Bulletin, 52. Meteoritics 9:101–121

Clarke RS Jr, Sellamuthu R, Goldstein JI (1984) Massive schreibersite in the Bellsbank and Santa Luzia meteorites. 47th annual meeting of the meteoritical society. Meteoritics 19:208

Connolly HC Jr, Zipfel J, Grossman JN, Folco L, Smith C, Jones R, Righter K, Zolensky M, Russell SS, Benedix G, Yamaguchi A, CohenBA (2006) The Meteoritical Bulletin, N°90, 2006 March. Meteorit Planet Sci 41(9):1383–1418

Connolly HC Jr, Zipfel J, Folco L, Smith C, Jones R, Benedix G, Righter K, Yamaguchi A, Chennaoui Aoudjehane H, Grossman JN (2007) The Meteoritical Bulletin, N°91, 2007 March. Meteorit Planet Sci 42(3):413–466

Connolly HC Jr, Smith C, Benedix G, Folco L, Righter K, Zipfel J, Yamaguchi A, Chennaoui Aoudjehane H (2008) The Meteoritical Bulletin, N°93, 2008 March. Meteorit Planet Sci 43(3):571–632

Crozaz G, McKay G (1990) Rare earth elements in Angra Dos Reis and Lewis Cliff 86010, two meteorites with similar but distinct magma evolutions. Earth Planet Sci Lett 97(3–4):369–381

Curvello WS (1950a) A preliminary note on the Casimiro de Abreu meteorite. Boletim do Museu Nacional, Rio de Janeiro, Geologia:11

Curvello WS (1950b) Metallographic study of the Cratheús Iron. Boletim do Museu Nacional, Rio del Janeiro, Geologia 10:120–127

Curvello WS (1950c) On a new fragment of the Santa Luzia de Goiás meteorite. Boletim do Museu Nacional, Rio del Janeiro, Geologia:9

Curvello WS (1954) Anais da. Academia Brasileira de Ciências 26:2

Curvello WS (1958) Boletim do Museu Nacional. Rio de Janeiro 27:23

Curvello WS, Ferreira CS (1951) Metallographic Study of the Barbacena Meteorite. Boletim du Museo Nacional, Rio del Janeiro, Geologia:14

Curvello WS, Ferreira CS (1952) The Pará de Minas Meteorite. Boletim du Museo Nacional, Rio del Janeiro, Geologia 18:1–10

Curvello WS, Malvin DJ, Wasson JT (1983) Bocaiuva: a unique silicate-inclusion-bearing iron meteorite. Meteoritics 18:285

Danon J, Scorzelli R, Souza Azevedo I, CurvelloW, Albertsen JF, Knudsen JM (1979) Iron-nickel 50-50 superstructure in the Santa Catharina meteorite. Nature 277:283–284
Davis AM, Dufek JD, Wadhwa M (2001) Euhedral phosphate grains in vugs and vesicles in ordinary chondrites, lunar samples and the ibitira eucrite: implications for trace element transport processes. Meteorit Planet Sci 36(Supplement):A47
Desnoyers C, Michel-Levy MC, Souza Azevedo I, Scorzelli RB, Danon J, da Silva GE (1985) Mineralogy of the Bocaiuva iron meteorite—a preliminary study. Meteoritics 20:113–124
Dreher AM, Dall'Agnol R, Martini SL (1995) The Ipitinga H_5 Chondrite: a new meteorite found in Pará State, Northern Brazil. Anais da Academia Brasileira de Ciências 67(1):45–54
Fireman EL, Schwarzer D (1957) Measurement of Li^6, He^3, and H^3 in meteorites and its relation to cosmic radiation. Geochim Cosmochim Acta 11(4):252–262
Fodor L, Keil K, Gomes CB (1977) Studies of Brazilian meteorites IV. Origin of a dark-colored unequilibrated lithic fragment in the Rio Negro chondrite. Revista Brasileira de Geociências 7(1):45–57
Frederick CL (1956) A classificational catalog of the meteoritic falls of the world. University of California publications in astronomy, vol 2 (1). University of California Press, Berkeley
Fredriksson K, Wlotzka F (1985) Morro do Rocio—an unequilibrated H_5 chondrite. Meteoritics 20:467–478
Funaki M, Danon J (1997) Natural remanent magnetization of Nova Petrópolis iron meteorite. Antarctic Meteorites XXII. 22nd symposium on antarctic meteorites, pp 43–45
Funaki M, Taguchi I, Danon J, Nagata T (1987) Magnetic and metallographical studies of Bocaiuva iron meteorite. Antarctic Meteorites XII. 12th symposium on antarctic meteorites, pp 132–133
Garvie LAJ (2012) The Meteoritical Bulletin, No 99, April 2012. MAPS 47(11):E1–E52
Goldstein JI, Reisner RJ, Rancourt DG, Lagarec K, Scorzelli RB (1998) The Santa Catharina meteorite: a cloudy zone microstructure consisting of a fine intergrowth of tetrataenite and antitaenite. 61st annual meteoritical society meeting
Gomes CB, Keil K (1977) Studies of Brazilian meteorites. Mineralogy and petrology of the Sete Lagoas, Minas Gerais, chondrite. Boletim IG-USP. Série Científica 7(1):77–82
Gomes CB, Keil K, Jarosewich E (1977a) Studies of Brazilian meteorites. VII. Mineralogy, petrology, and chemistry of the Uberaba, Minas Gerais, chondrite. Anais Academia Brasileira de Ciencias 49:269–274
Gomes CB, Keil K, Jarosewich E, Curvello WS (1977b) Studies of Brazilian meteorites VIII. Mineralogy, petrology and chemistry of the Itapicurú Mirim, Maranhao, chondrite. Anais de Academia Brasileira de Ciencias 49:407–412
Gomes CB, Jarosewich E, Keil K, Curvello WS (1977c) Studies of Brazilian meteorites. IX. Mineralogy, petrology and chemistry of the Macau, Rio Grande do Norte, chondrite. Anais de Academia Brasileira de Ciencias 49:575–579
Gomes CB, Keil K, Ruberti E, Jarosewich Silva JMLU (1978a) O meteorito Ipiranga, município de Foz do Iguaçu, Paraná: características mineralógicas, petrológicas e químicas. XXX Congresso Brasileiro Geologia 1:98–99
Gomes CB, Keil K, Ruberti E, Jarosewich Silva JMLU (1978b) Studies of Brazilian Meteorites XVI. Mineralogy, petrology and chemistry of the Ipiranga, Paraná, chondrite. Chemie der Erde 37:265–270
Graham AL (1980) The meteoritical bulletin. Meteoritics 58:235–240
Graham AL (1981) The Meteoritical bulletin, 59. Meteoritics 16:193–199
Graham AL (1984) The meteoritical bulletin, 62. Meteoritics 19:49–57
Graham AL (1988) The meteoritical bulletin, 66. Meteoritics 23:171–173
Graham AL (1990) The meteoritical bulletin, 68. Meteoritics 25:59–70
Grossman JN (1998) The meteoritical bulletin, No 82*, 1998 July. Meteorit Planet Sci 33:A221–A239
Grossman JN (1999) The meteoritical bulletin, No 83*, 1999 July. Meteorit Planet Sci 34:A169–A186
Grossman JN (2000) The meteoritical bulletin, No 84*, 2000 August. Meteorit Planet Sci 35:A199–A225
Grunewaldt H (1983) Anais Academia Brasileira Ciencias 55:1

Harlow GE, Prinz M, Nehru CE, Taylor GJ, Keil K (1977) Pyroxene Relations in the Serra de Magé Meteorite. Meteoritics 12:252

Harlow GE, Nehru CE, Prinz M, Taylor GJ, Keil K (1979) Pyroxenes in Serra de Magé - Cooling history in comparison with Moama and Moore County. Earth Planet Sci Lett 43(2):173–181

Harvey RP, McSween HY Jr (1991) New Observations of Nakhla, Governador Valadares and Lafayette, and Their Bearing on Petrogenesis. Abstracts of the Lunar and Planetary Science Conference vol 22, p 527

Jago RA (1979) Santa Catharina and the origin of cloudy taenite in meteorites. Nature 279(5712):413–415

Johnson CA, Prinz M, Weisberg MK (1989) Magnetite in silicate inclusions in the Bocaiuva iron: formation by reaction of CO_3 chondritic silicates and metal. Meteoritics 24:281

Keil K, Kirchner E, Gomes CB, Nelen J (1977) Studies of Brazilian meteorites V. Evidences for shock metamorphism in the Paranaíba, Mato Grosso, chondrite. Revista Brasileira de Geociências 7(3):256–268

Keil K, Kirchner E, Gomes CB, Jarosewich Murta RLL (1978) Studies of Brazilian meteorites XIV. Mineralogy, petrology and chemistry of the Conquista, Minas Gerais, chondrite. Meteoritics 13:177–187

Keil K, Lange D, Ulbrich MNC, Gomes CB, Jarosewich E, Roisenberg A, Souza MJ (1978b) Studies of Brazilian meteorites XIII. Mineralogy, petrology and chemistry of the Putinga, Rio Grande do Sul, chondrite. Meteoritics 13:165–175

Kessel R, Beckett JR, Huss GR, Stolper EM (2004) The activity of chromite in multicomponent spinels: Implications for T-fO_2 conditions of equilibrated H chondrites. Meteorit Planet Sci 39(8):1287–1305

Kohout T, Pesonen LJ (2005) Chondrule magnetic conglomerate test of Avanhandava H_4 chondrite. 68th annual meteoritical society meeting. Gatlinburg, Tennessee (USA). Meteorit Planet Sci Supplement:5202

Kohout T, Kletetschka G, Pesonen LJ, Wasilewski PJ (2006) Magnetic studies of Avanhandava H_4 and Bjurböle L_4 chondrules. 37th annual lunar and planetary science conference, League City, Texas (USA), p 1601

Kohout T, KletetschkaG, Pesonen LJ (2007) Identification of the shock effects in the the Avanhandava H_4 Chondrules based on the coercivity spectra of the remanent magnetization. 38th lunar and planetary science conference, (Lunar and Planetary Science XXXVIII), League City, Texas. LPI 1338, p 1773

Korochantseva EV, Schwenzer SP, Buikin AI, Hopp J, Ott U, Trieloff M (2011) ^{40}Ar-^{39}Ar and cosmic-ray exposure ages of nakhlites -Nakhla, Lafayette, Governador Valadares- and Chassigny. Meteorit Planet Sci 46(9):1397–1417

Kracher A, Willis J, Wasson JT (1980) Chemical classification of iron meteorites; IX. A new group (IIF), revision of IAB and IIICD, and data on 57 additional irons. Geochimica Cosmochimica Acta 44:773–787

Kuhn IA (2008) Microanálise quantitativa por EDS/MEV das fases metálicas dos meteoritos Putinga e Vaca Muerta. Salão de Iniciação Científica (20. 2008 out. 20-24: Porto Alegre, RS). Livro de resumos. Porto Alegre: UFRGS

Lange DE, Keil K, Gomes CB (1979) The Mafra meteorite and its lithic clasts: a genomict L-group chondrite breccia. Meteoritics 14:472

Levi-Donati GR, Shima M, Sighinolfi GP (1976) Brazilian meteorites—The Mafra, Santa Catarina State, chondrite. Meteoritics 11:29–41

Liu M, Scott ERD, Keil K, Wasson JT, Clayton RN, Mayeda T, Eugster O, Crozaz G, Floss C (2001) Northwest Africa 176: a unique iron meteorite with silicate inclusions related to Bocaiuva. 32nd annual lunar and planetary science conference, Houston, Texas (USA), p 2152

Lofgren GE, Lanier AB (1992) Dynamic crystallization experiments on the Angra Dos Reis achondritic meteorite. Earth Planet Sci Lett 111(2–4):455–466

Lux G, Keil K, Taylor GJ (1981) Chondrules in H3 chondrites—textures, compositions and origins. Geochim Cosmochim Acta 45:675–685

Ma MS, Murali AV, Schmitt RA (1976) Genesis of the Angra dos Reis and other achondritic meteorites. Meteoritics 11:323

Malvin DJ, Wasson JT, Clayton RN, Mayeda TK, Curvello WDS (1985) Bocaiuva—a silicate-inclusion bearing iron meteorite related to the Eagle-Station pallasites. Meteoritics 20:259–273

Meen VB (1939) Santa Luzia de Goias meteorite. Am Mineral 9

Menezes V (1957) A probable meteorite fall in Brazil. Sky and telescope XVII(1):10

Mikouchi T, Monkawa A, Tachikawa O, Yamada I, Komatsu M, Koizumi E, Chokai J, Miyamoto M (2004) Electron backscatter diffraction and forescatter electron image analyses of the Governador Valadares nakhlite. Meteoritics & Planetary Science, vol 39. In: Proceedings of the 67th annual meeting of the meteoritical society, Rio de Janeiro, Brazil, Supplement, A69

Miller MK, Russell KF (1992) An APFIM investigation of a weathered region of the Santa Catharina meteorite. Surf Sci 266:441–445

Mittlefehldt D (2005) Ibitira: a basaltic achondrite from a distinct parent asteroid and implications for the Dawn mission. Meteorit Planet Sci 40:665

Mittlefehldt DW, Lindstrom MM (1990) Geochemistry and genesis of the angrites. Geochim Cosmochim Acta 54:3209–3218

Mittlefehldt DW, Lindstrom MM (1996) Martian Meteorites QUE 94201: an unusual basalt, and Governador Valadares, a typical clinopyroxenite: geochemistry. Lunar Planet Sci 27:887

Miyamoto M (2004) Electron backscatter diffraction and forescatter electron image analyses of the Governador Valadares Nakhlite. Meteorit Planet Sci 39(Supplement):5128

Nunes GA, da Costa AR, Cassino FSL, Lays FS, de Souza PA (2010a) Microstructural characterization of a probable Itutinga meteorite fragment. Revista Escola de Minas 63(3):425–431

Nunes GA, da Costa AR, Cassino FSL, de Souza PA (2010b) Microstructural characterization of a probable Itutinga meteorite fragment. 73rd annual meeting of the meteoritical society, New York (USA). Meteorit Planet Sci Supplement:5068

Nyquist LE, Bansal B, Wiesmann H, Shih CY (1994) Neodymium, strontium and chromium isotopic studies of the LEW86010 and Angra Dos Reis meteorites and the chronology of the angrite parent body. Meteoritics 29(6):872–885

Oliveira de JCP, Costa da MI Jr, Vasquez A, Roisenberg A, Vieira N Jr (1988) Moessbauer study of the Putinga chondrite. Physica Scripta37(1):185–187

Paar W, Keil K, Gomes CB, JarosewichE (1976) Studies of Brazilian meteorites II. The Avanhandava chondrite: mineralogy, petrology and chemistry. Revista Brasileira de Geociências 6(3):201–210

Paduani C, Samudio Pérez CA, Ardisson JD (2005) A mössbauer effect study of the soledade meteorite. Brazilian J Phys 35:667–669

Pinheiro GMS, Acevedo RD, Pimentel MM (2006) Caracterização Petrográfica do condrito ordinário de Santa Vitória do Palmar, RS, Brasil. XLIII Congresso Brasileiro de Geologia, Aracaju-SE. Anais do XLIII Congresso Brasileiro de Geologia. SBG Núcleo Bahia-Sergipe 1:318

Prinz M, Keil K, Hlava PF, Berkley JL, Gomes CB, Curvello WS (1977) Studies of Brazilian meteorites III. Origin and history of the Angra dos Reis achondrite. Earth Planet Sci Lett 35:317–330

Pucheta FN, Cassino FSL (2008) Brazilian meteorite Patos de Minas (octahedrite). 71st annual meeting of the meteoritical society, Matsue (Japan). Meteorit Planet Sci 43 (Abstracts):5116

Rambaldi ER, Waenke H, Larimer JW (1979) Interelement refractory siderophile fractionation in ordinary chondrites. In: Proceedings of the lunar and planetary science conference, vol 1, pp 997–1010. Pergamon Press, Inc., New York

Riches AJV, Day JMD, Walker RJ, Simonetti A, Liu Y, Neal CR, Taylor LA (2012) Rhenium-osmium isotope and highly-siderophile-element abundance systematics of angrite meteorites. Earth Planet Sci Lett 353:208–218

Rios DC, Carvalho WP (2009) Bendegó: um visitante do espaço. Editora Cedraz, Salvador 16 p

Romig AD, Goldstein JI (1981) Low temperature phase equilibria in the Fe-Ni and Fe-Ni-P systems—application to the thermal history of metallic phases in meteorites. Geochim Cosmochim Acta 45:1187–1197

Russell SS, Zipfel J, Grossman JN, Grady MM (2002) The meteoritical bulletin, No 86. Meteorit Planet Sci 37(Supplement):A157–A184

Scorzelli RB, DanonJ (1986) Mössbauer Study of Schreibersite from Bocaiuva Iron Meteorite. Abstracts and Program for the 49th annual meeting of the meteoritical society, New York (USA), vol 600, p 29

Scorzelli RB, Souza Azevedo I, ZucolottoME (2003) Revisiting the Brazilian Hexahedrites: Angra Dos Reis (Iron), Pirapora, and Patos De Minas. Meteoritics & Planetary Science, vol 38, Supplement, Abstract 5017, Meteoritical Bulletin 8

Sears PM (1963) Recovery of the Bendegó Meteorite. Meteoritics 2(1):22

Shima M, Sighinolfi GP, Jochum KP, Hintenberger H (1973) The Parambu Meteorite: bulk chemistry and heavy trace metals by spark mass spectroscopy (abstract). Meteoritics 8:440–441

Shima M, Jochum KP, Sighinolfi GP, Hintenberger H (1974) The chemical composition of major elements and heavy trace metals in chondrites Parambu and Marília. Meteoritics 9:199–207

Sighinolfi GP, Gorgoni C (1983) Lithophile elements in individual chondrules of the H_4 Mafra chondrite 18:197–208

Souza Azevedo I, Scorzelli RB, Danon J, Zucolotto ME (1987) Mössbauer Spectroscopy Study of the Nova Petropoils Meteorite and Its Corrosion Mechamism. Meteoritics 22:320

Spencer LJ, Hey MH (1930) A New Meteoric Iron from Piedade Do Bagre, Minas Geraes, Brazil. Mineral Mag 22:271–282

Steele IM, Smith JV (1976) Mineralogy of the Ibitira eucrite and comparison with other eucrites and lunar samples. Earth Planet Sci Lett 33(1):67–78

Svisero DP, Amaral SE, Gomes CB (1980) O meteorito Itapuranga, Goiás, e os sideritos brasileiros. Boletim IG, Instituto de Geociências, USP 2:21–30

Symes RF, Hutchinson R (1970) Medanitas and Putinga, two South American meteorites. Miner Mag 37:721–723

Treiman AH (1988) Angra Dos Reis is not a cumulate igneous rock. Abstracts of the Lunar and Planetary Science Conference, vol 19, p 1203

Treiman AH, Jones JH, JanssensMJ, Wolf R, Ebihara M (1988) Angra Dos Reis: Complex Silicate Fractionations. 51st annual meeting of the meteoritical society. Fayetteville, Arkansas (USA). Lunar and Planetary Institute, vol 665, p 156

Treiman AH, Lanzirotti A, Xirouchakis D (2004) Ancient water on asteroid 4 Vesta: evidence from a quartz veinlet in the Serra de Magé eucrite meteorite. Earth Planet Sci Lett 219(3–4):189–199

Van Tassel R, Dillen H, Vochten R, De Grave E, Hertogen J (1992) An overlooked fragment of the Santa Catharina ataxite. Meteoritics 27(4):467–469

Vianna Cautinho JM (1999) Quijingue, Bahia, the First Brazilian Pallasite. Meteorit Planet Sci 35(Supplement):A179

Wang PL, Rumble D, McCoy TJ (2001) Oxygen isotopic compositions of IVA iron meteorite: new views from in situ UV laser microprobe analyses 64th Annual Meeting of the Meteoritical Society, Vatican City. Meteorit Planet Sci 36(Supplement):5243

Wasserburg GJ, Tera F, Papanastassiou y DA, Huneke JC (1977) Isotopic and chemical investigation of Angra dos Reis. Earth Planet Sci Lett 35:294 –316

Weisberg MK, Smith C, Benedix G, Folco L, Righter K, Zipfel J, Yamaguchi A, Chennaoui Aoudjehane H (2009) The meteoritical bulletin, N° 95, 2007 March. Meteorit Planet Sci 44(3):1–33

Weisberg MK, Smith C, Benedix G, Herd CDK, Righter K, Haack H, Yamaguchi A, Chennaoui Aoudjehane H, Grossman JF (2009) The meteoritical bulletin, N° 96, September 2009. Meteorit Planet Sci 44(9):1355–1397

Wilkening LL, Anders E (1974) Some studies of the unusual eucrite, Ibitira. Meteoritics 9:422

Wlotzka F (1985) Olivine-spinel and olivine-ilmenite thermometry in chondrites of different petrologic type. Lunar Planet Sci 16:918–919

Wlotzka F (1992) The meteoritical bulletin, No 72*. Meteoritics 27(109–117):477–483

Wlotzka F (1994) The meteoritical bulletin, No 77*, 1994 November. Meteoritics 29:891–897

Wlotzka F (1995) The meteoritical bulletin, No 78*, 1995 November. Meteoritics 30:792–796

Wlotzka F, Fredriksson K (1980) Morro do Rocio, an unequilibrated H_5 chondrite. 43rd annual meeting of the meteoritical society, La Jolla, California (USA), Lunar and Planetary Institute Contribution, vol 412, p 124

Yanai K (1994) Comparative studies of three angrites; Angra dos Reis, LEW87051 and Asuka-881371 meteorites. Antarctic Meteorites XIX. 19th symposium on antarctic meteorites, NIPR, Tokyo, pp 51–54

Zanardo A, Navarro GRB, del Roveri C, Morales N (2011) The chondritic meteorite Saulo Gomes. Geociências (São Paulo) 30(2)

Zucolotto EM (1999) Brazilian meteorites. Meteorite 5:8–11

Zucolotto ME (2004) Brazilian meteorites. Meteorit Planet Sci Tucson, Planetary Society 39:A119

Zucolotto ME, Antonello LL (2001) Mineralogy and petrography of the L6 Chondrite Rio do Pires, Brazil. Meteorit Planet Sci 36(9), Supplement:A234

Zucolotto ME, Antonello LL (2004a) Brazilian meteorites. 67th annual meeting of the meteoritical society, Rio de Janeiro (Brazil). Meteorit Planet Sci 39 (Supplement):5089

Zucolotto ME, Antonello LL (2008) Studies of Brazilian Meteorites of the Museu Nacional - I: Petrography and Mineralogy of the Santa Vitória do Palmar, Rio Grande do Sul, L_3 Chondrite Meteorite. Arquivos do Museu Nacional 66:611–629

Zucolotto EM, Carvalho WP (2009) The Paranaiba (Cacilandia) fireball, MS, Brazil. 72th annual meeting of the meteoritical society (Nancy), France, Meteorit Planet Sci 44 (Supplement):5448

Zucolotto ME, Monteiro F (2012) A new mass and the history of São João Nepomuceno. 75th annual meeting of the meteoritical society, Cairns (Australia). Meteorit Planet Sci 47 (Supplement):5152

Zucolotto EM, Pinto AL (2000) Electron back-scattered diffraction studies of the Barbacena meteorite. Meteorit Planet Sci 36(Supplement):A180

Zucolotto EM, Riff D (2009) Two new meteorites from Bahia, Brazil. 72th annual meeting of the meteoritical society, Nancy (France). Meteorit Planet Sci 44 (Supplement):5380

Zucolotto EM, de Carvalho W, Gomes SO (1999) The Bendegó Iron. Meteorite 5(4):36–39

Zucolotto ME, Antonello LL, Scorzelli RB, Munayco P, dos Santos E, Varela ME, Roisemberg A (2010a) Petrography and mineralogy of Lavras Do Sul Meteorite. 73th annual meeting of the meteoritical society, New York (USA). Meteorit Planet Sci 45 (Supplement):5036

Zucolotto ME, Martini P, Monzon JM, Varela ME, Antonello O (2010b) A fireball and at least three distinct meteorites on the border betwee Brazil and Uruguay. 73th annual meeting of the meteoritical society, New York (USA). Meteorit Planet Sci 45 (Supplement):5162

Zucolotto ME, Antonello LL, Scorzelli RB, Varela ME, Magnelli DE, Munayco P, dos Santos E, LudkaIP (2011a) Varre-Sai (L_5)-A recent meteorite fall from Brazil. 74th annual meeting of the meteoritical society, London. Meteorit Planet Sci 46 (Supplement):5119

Zucolotto ME, Grillo O, Antonello LL (2011b) Mechanical disassembling and tessellation reassembling, tolos for understanding the Widmanstatten structure. 74th annual meeting of the meteoritical society, London. Meteorit Planet Sci 46 (Supplement):5214

Zucolotto ME, Antonello LL, Varela ME, Scorzelli RB, Ludka IP, Munayco P, dos Santos E (2012a) Lavras do Sul: a new equilibrated ordinary L_5 Chondrite from Rio Grande do Sul, Brazil. Earth Moon Planet 108(2):139–150

Zucolotto ME, Antonello LL, Varela ME, Scorzelli RB, Munayco P, dos Santos E, Ludka IP (2012b) Varre-Sai: The Recent Brazilian Fall. Earth Moon Planet 109:43–53

Chapter 5
Chile

Although meteorites can fall anywhere, it is easier to find them over deflation surfaces that are geomorphological critical localities situated in desert of Atacama (Philippi 1856, 1860; Domeyko 1862, 1864a, b, 1875; Fletcher 1889; Scherer and Delisle 1992; Muñoz et al. 2007) in Fig. 5.1. Alone in the first half of this year (2013) 238 new meteorites were classified for this country. All of them were picked up from the Atacama's desert deflation surfaces.

There are no discoveries of meteorites south of Santiago de Chile.

The info of all meteorites reported until June 30th 2013 is given below.

Algarrobo. 27° 5′ S, 70° 35′ W. Iron, fine octahedrite, (IAB ungrouped). Find, 1959, 1,280 g. IGPP-UCLA. Wlotzka (1991). Fig. 5.2.

Baquedano. 23° 18′ S, 69° 53′ W. Antofagasta, Iron, (IIIAB). Find, 1932, 22 kg. Natural History Museum. Fig. 5.3.

Barranca Blanca. 28° 05′ S, 69° 20′ W. Atacama. Iron, (IIE anomalous). Find, 1855, 12 kg. Natural History Museum. Scott and Wasson (1976), Wasson and Wang (1986).

Blanca Estela. 25° 00′ S, 69° 30′ W. Antofagasta. Iron, (IAB complex). Ni = 6.6 wt%. Find, 2002, 15.6 kg. Connoly et al. (2008).

Cachiyuyal. 25° 00′ S, 69° 30′ W. Antofagasta. Iron, (IIIE). Find, 1874, 2.5 kg. Natural History Museum.

Caldera. 27° 03′ S, 70° 48′ W. Atacama. Achondrite, eucrite. Find, 1967, 500 g. Smithsonian Institution collections. Wlotzka (1993). Fig. 5.4.

Caleta el Cobre. 24° 15′ S, 70° 31′ W (Caleta el Cobre 001, 24° 16′ S, 70° 02′ W). Antofagasta. Stone, ordinary chondrites. Finds. 20 meteorites so far. Table 5.1. Fig. 5.5.

Carcote. 24° S, 69° W. Antofagasta. Stone, ordinary chondrite, (H_5). Find, 1888, 392 g. Natural History Museum.

Catalina. 25° 14′ S, 69° 43′ W (Catalina 003 25° 12′ 16.8″ S, 69° 49′ 42.8″ W., Catalina 037 25° 05′ 54″ S, 69° 45′ W). Antofagasta. Finds. 36 meteorites so far. Table 5.2. Figs. 5.6 and 5.7.

R. D. Acevedo et al., *Catalogue of Meteorites from South America*, SpringerBriefs in Earth System Sciences, DOI: 10.1007/978-3-319-01925-3_5,

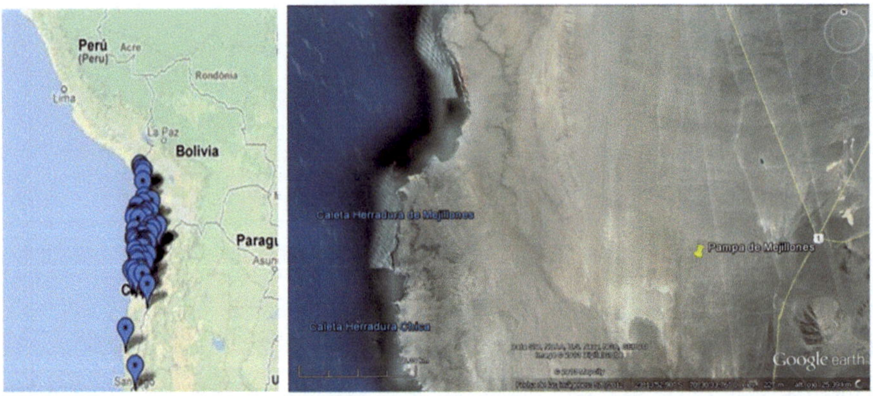

Fig. 5.1 Chilean meteorites are concentrated in deflation surfaces of Antofagasta and Atacama. *Credit*: Google Maps Imágenes © 2013 TerraMetrics Datos de mapa © 2013 Google Inav/geosistemas SRL, Google Earth © 2013 Mapcity

Fig. 5.2 Algarrobo. *Credit*: Unknown

Cerro del Inca. 22° 13′ 0″ S, 68° 54′ 30″ W. Antofagasta. Iron, octahedrite, (IIIF). Ni = 7.69 wt%. Find, 1997, 20.6 kg. IGPP-UCLA. Grossman (2000). Fig. 5.8.

Chañaral. 26° 30′ S, 70° 15′ W. Atacama. Iron, (IIIAB). Find, 1884, 1,207 g.

Cobija. 22° 34′ S, 70° 15′ W. Antofagasta. Stone, ordinary chondrite, (H6). Find, 1892, 6.53 kg. "Lampa" in Natural History Museum collection. Fig. 5.9.

Copiapó. 27° 18′ S, 70° 24′ W. Atacama. Iron, (IAB complex). Find, 1863, 20 kg. Natural History Museum. Fig. 5.10.

Fig. 5.3 Baquedano. *Credit*: Jay Piatek

Fig. 5.4 Caldera. *Credit*: Corey Kuo

Corrizatillo. 26° 02′ S, 70° 20′ W. Atacama. Iron, (IIICD) (Muñoz et al. 2007) or (IAB complex) (MetSoc Database). Find, 1884, 1,328 g.

Dehesa. 33° 30′ S, 70° 30′ W. Región Metropolitana. Iron, ungrouped. Find, 1866, 280 g. Natural History Museum.

Dolores. 19° 39′ S, 69° 57′ W. Tarapacá. Iron, (IIIAB). Ni = 7.48 wt%. Find, 2001, 1,328 g. IGPP-UCLA. Russell et al. (2004). Fig. 5.11.

El Médano. 24° 51′ S, 70° 32′ W. Antofagasta. Finds. 196 meteorites so far. Table 5.3. Fig. 5.12.

Elqui. Unknown coordenates. Iron, hexaedrite, (IIAB). Ni = 5.96 wt%. Find, 1990, 260 g. University of La Serena (Chile), IGPP-UCLA. Grossman (1998).

Estación Imilac. 24° 13′ 47″ S, 68° 53′ 33″ W. Antofagasta. Stone, ordinary chondrite, (H_5), S_4 W_1. Find, 2004, 1.9 g. University of Chile, Johnson Space Center, Smithsonian Institution collections.

Guanaco. 25° 6′ S, 69° 32′ W. Antofagasta. Iron, (IIG). Ni 44.3 mg/g. Find, 2000, 51.7 kg. IGPP-UCLA. Russell et al. (2004). Fig. 5.13.

Ilimaes (hierro). 26° S, 70° W Atacama. Iron, (IIIAB). Find, 1870, 51.7 kg. Natural History Museum.

Table 5.1 Caleta el Cobre

Name	Class	W	Fa	Fs	Wo	χ	Found	Weight (g)	Curator	References
Caleta el Cobre 001	H_6	W_4	18.41 ± 0.13	16.25 ± 0.39	1.47 ± 0.2	4.5	2010	9	CEREGE, U. Chile	Met. Bull., No. 100, MAPS 46, in preparation (2013)
Caleta el Cobre 002	H_6	W_3	19.08 ± 0.41	16.81 ± 0.33	1.36 ± 0.22	4.55	2010	18	CEREGE, U. Chile	Met. Bull., No. 100, MAPS 46, in preparation (2013)
Caleta el Cobre 003	H_5	W_2	18.54 ± 0.17	16.77 ± 1.15	1.19 ± 0.19	5.08	2010	338	CEREGE, U. Chile	Met. Bull., No. 100, MAPS 46, in preparation (2013)
Caleta el Cobre 004	H_5	W_2	18.05 ± 0.28	15.88 ± 0.29	1.06 ± 0.22	5.25	2010	254	CEREGE, U. Chile	Met. Bull., No. 100, MAPS 46, in preparation (2013)
Caleta el Cobre 005	H_5	W_2	18.64 ± 0.80	16.96 ± 1.14	0.99 ± 0.39	5.04	2010	193	CEREGE, U. Chile	Met. Bull., No. 100, MAPS 46, in preparation (2013)
Caleta el Cobre 006	L_6	W_3	25.43 ± 1.03	20.59 ± 0.43	1.53 ± 0.31	4.53	2010	179	CEREGE, U. Chile	Met. Bull., No. 100, MAPS 46, in preparation (2013)
Caleta el Cobre 007	L_6	W_3	22.97 ± 0.27	19.44 ± 0.42	0.82 ± 0.33	4.52	2010	24	CEREGE, U. Chile	Met. Bull., No. 100, MAPS 46, in preparation (2013)
Caleta el Cobre 008	H_6	W_3	19.57 ± 0.73	17.04 ± 1.10	1.12 ± 0.51	4.62	2010	15	CEREGE, U. Chile	Met. Bull., No. 100, MAPS 46, in preparation (2013)
Caleta el Cobre 009	L_4	W_1	22.73 ± 0.35	19.05 ± 0.22	0.99 ± 0.22	4.77	2010	95	CEREGE, U. Chile	Met. Bull., No. 100, MAPS 46, in preparation (2013)
Caleta el Cobre 010	L_4	W_1	22.99 ± 0.52			4.90	2010	47	CEREGE, U. Chile	Met. Bull., No. 100, MAPS 46, in preparation (2013)
Caleta el Cobre 011	L_6	W_4	25.07 ± 0.37	20.52 ± 0.19	1.25 ± 0.23	4.39	2010	32	CEREGE, U. Chile	Met. Bull., No. 100, MAPS 46, in preparation (2013)
Caleta el Cobre 012	L_4	W_1	22.89 ± 0.60	19.56 ± 0.29	1.11 ± 0.15	4.81	2010	17	CEREGE, U. Chile	Met. Bull., No. 100, MAPS 46, in preparation (2013)
Caleta el Cobre 013	L_4	W_3	22.98 ± 1.14	19.45 ± 0.15	1.09 ± 0.12	4.73	2010	105	CEREGE, U. Chile	Met. Bull., No. 100, MAPS 46, in preparation (2013)

(continued)

Table 5.1 (continued)

Name	Class	W	Fa	Fs	Wo	χ	Found	Weight (g)	Curator	References
Caleta el Cobre 014	H_4	W_2	18.75 ± 0.13	16.45 ± 0.33	1.19 ± 0.37	4.82	2010	20	CEREGE, U. Chile	Met. Bull., No. 100, MAPS 46, in preparation (2013)
Caleta el Cobre 015	L_6	W_3	25.36 ± 0.66	20.82 ± 0.26	1.74 ± 0.13	4.34	2010	115	CEREGE, U. Chile	Met. Bull., No. 100, MAPS 46, in preparation (2013)
Caleta el Cobre 016	H_5	W_2	18.58 ± 0.51	16.64	1.25	4.83	2010	34	CEREGE, U. Chile	Met. Bull., No. 100, MAPS 46, in preparation (2013)
Caleta el Cobre 017	H_5	W_2	19.43 ± 0.95	16.59 ± 0.30	1.31 ± 0.15	4.95	2010	41	CEREGE, U. Chile	Met. Bull., No. 100, MAPS 46, in preparation (2013)
Caleta el Cobre 018	H_5	W_4	18.1 ± 0.5	14.9	0.8	4.47	2010	6	CEREGE, U. Chile	Met. Bull., No. 100, MAPS 46, in preparation (2013)
Caleta el Cobre 019	H_6	W_3	18.8 ± 0.3	16.2 ± 0.3	1.4 ± 0.1	4.62	2010	1.1	CEREGE, U. Chile	Met. Bull., No. 100, MAPS 46, in preparation (2013)
Caleta el Cobre 020	H_5	W_3	18.7 ± 0.3	16.6 ± 0.2	1.6 ± 0.1	4.73	2011	633	CEREGE	Met. Bull., No. 102, MAPS 48, in preparation (2014)

University of Chile (U. Ch); Centre de Recherche et d'Enseignement de Géosciences de l'Environnement (CEREGE)

Fig. 5.5 Caleta El Cobre. *Credit*: Egon Lobo Miranda

Imilac. 24° 12′ 12″ S, 68° 48′ 24″ W. Antofagasta. Stony-Iron, pallasite. Find, 1822, 920 kg. Natural History Museum. Pedersen and García (1987), Killgore (1997). Fig. 5.14.

Iquique. 20° 11′ S, 69° 44′ W. Tarapaca. Iron, (IVB). Find, 1871, 12.5 kg. Natural History Museum. Fig. 5.15.

Joel's Iron. 24° S, 69° W. Antofagasta. Iron, (IIIAB). Find, 1858, 1,300 g. Natural History Museum.

Juncal. 26° 0′ S, 69° 15′ W. Atacama. Iron, (IIIAB). Find, 1866, 104 kg. Natural History Museum.

La Primitiva. 19° 55′ S, 69° 49′ W. Tarapaca. Iron, (IIG). Find, 1888, 27.4 kg. Natural History Museum. Fig. 5.16.

La Serena. Unknown coordenates. Iron, octahedrite (IIICD). Ni = 7.62 wt%. Find, 1990, 663 g. University of La Serena, IGPP-UCLA. Grossman (1998).

La Yesera 001. 223° 16′ 14″ S, 70° 28′ 59″ W. Antofagasta. Stone, ordinary chondrite, (H_6), S_2 W_3. Find, 2003, 205 g. Russell et al. (2004). Johnson Space Center, Smithsonian Institution collections. Fig. 5.17.

La Yesera 002. 23° 16′ 14″ S, 70° 28′ 59″ W. Antofagasta. Stone, ordinary chondrite, (LL_5), S_2 W_2. Find, 2003, 2.63 kg. Russell et al. (2004). Johnson Space Center, Smithsonian Institution collections. Fig. 5.18.

La Yesera 003. 23° 17′ 27″ S, 70° 28′ 21″ W. Antofagasta. Stone, ordinary chondrite, (L_4), S_3 W_4. Find, 2003, 447 g. Johnson Space Center, Smithsonian Institution collections. Meteoritical Bulletin, no. 100, MAPS 46, in preparation (2013).

La Yesera 004. 23° 17′ 19″ S, 70° 28′ 29″ W. Antofagasta. Stone, ordinary chondrite, (L_6), S_2 W_3. Find, 2003, 1,489 g. University of Chile, Smithsonian Institution collections. Meteoritical Bulletin, no. 100, MAPS 46, in preparation (2013).

Las Cruces. 23° 22′ 59″ S, 70° 35′ 10″ W. Antofagasta. Iron (IIIAB). Ni, 20.6 μg/g. Find, 2001, 528 g. IGPP-UCLA. Meteoritical Bulletin, no. 101, MAPS 47, in preparation (2013). Fig. 5.19.

Table 5.2 Catalina

Name	Group	Class	W	S	Fa	Fs	Wo	Ni	χ	Found	Weight	Curator	References
Catalina 002	Stone, ordinary chondrite	LL_3	W_4		18	17	1.4		3.60	2010	61.1 g	CEREGE	Met. Bull., No. 101, MAPS 47, in preparation (2013)
Catalina 003	Iron, ataxite	IVB						0.17 μg/g		1999	3.18 kg	UCLA	Met. Bull., No. 101, MAPS 47, in preparation (2013)
Catalina 004	Stony-Iron, mesosiderite	A	Mod			35.7 ± 1.4	3.3 ± 0.4		5.62	2010	37.5 g	CEREGE, MM	Met. Bull., No. 101, MAPS 47, in preparation (2013)
Catalina 005	Stone, ordinary chondrite	H_4	W1		18.3 ± 0.2	16.5 ± 0.4	1.4 ± 0.2		5.24	2010	228 g	CEREGE	Met. Bull., No. 102, MAPS 48, in preparation (2014)
Catalina 006	Stone, ordinary chondrite	$H_{5/6}$	W2		19.6 ± 0.3	17.3 ± 0.5	1.9 ± 0.1		5.08	2010	19.5 g	CEREGE	Met. Bull., No. 102, MAPS 48, in preparation (2014)
Catalina 007	Stone, ordinary chondrite	H_4	W_2		18.7 ± 0.4	16.7 ± 0.9	1.3 ± 0.1		5.05	2010	11.9 g	CEREGE	Met. Bull., No. 102, MAPS 48, in preparation (2014)
Catalina 008	Carbonaceous chondrite	CO_3	Mod		19.1 ± 13.5	2.9 ± 1.0	5.5 ± 1.1.6		4.05	2011	98 g	CEREGE, MM	Met. Bull., No. 102, MAPS 48, in preparation (2014)
Catalina 009	Carbonaceous chondrite	CR_2	Mod		1.3–31.8	2.4	0.6		4.86	2012	5.2 g	CEREGE	Met. Bull., No. 102, MAPS 48, in preparation (2014)
Catalina 010	Stone, ordinary chondrite	L_5	W_2						4.59	2010	329 g	CEREGE	Met. Bull., No. 102, MAPS 48, in preparation (2014)

(continued)

Table 5.2 (continued)

Name	Group	Class	W	S	Fa	Fs	Wo	Ni	χ	Found	Weight	Curator	References
Catalina 011	Stone, ordinary chondrite	H_5	W_2						5.10	2010	573 g	CEREGE	Met. Bull., No. 102, MAPS 48, in preparation (2014)
Catalina 012	Stone, ordinary chondrite	H_6	W_2			18.0	0.8		4.75	2010	225 g	CEREGE	Met. Bull., No. 102, MAPS 48, in preparation (2014)
Catalina 013	Stone, ordinary chondrite	H_4	W_1		18.0 ± 0.1	17.5 ± 2.3	1.1 ± 0.5		5.13	2010	428 g	CEREGE	Met. Bull., No. 102, MAPS 48, in preparation (2014)
Catalina 014	Stone, ordinary chondrite	H_4	W_1		19.3 ± 0.9	17.9 ± 1.3	1.0 ± 0.8		5.19	2010	191 g	CEREGE	Met. Bull., No. 102, MAPS 48, in preparation (2014)
Catalina 015	Stone, ordinary chondrite	L_5	W_1			19.8	1.9		4.90	2009	239 g	CEREGE	Met. Bull., No. 102, MAPS 48, in preparation (2014)
Catalina 016	Stone, ordinary chondrite	H_4	W_2						5.20	2010	647 g	CEREGE	Met. Bull., No. 102, MAPS 48, in preparation (2014)
Catalina 017	Stone, ordinary chondrite	H_5	W_1						5.23	2010	426 g	CEREGE	Met. Bull., No. 102, MAPS 48, in preparation (2014)
Catalina 018	Stone, ordinary chondrite	L_6	W_2						4.55	2010	1,018 g	CEREGE, MM	Met. Bull., No. 102, MAPS 48, in preparation (2014)
Catalina 019	Stone, ordinary chondrite	H_4	W_2		18.4 ± 0.5	17.2 ± 0.5	1.4 ± 0.9		5.10	2010	3.19 kg	CEREGE, MM	Met. Bull., No. 102, MAPS 48, in preparation (2014)

(continued)

Table 5.2 (continued)

Name	Group	Class	W	S	Fa	Fs	Wo	Ni	χ	Found	Weight	Curator	References
Catalina 020	Stone, ordinary chondrite	L_6	W_2						4.63	2010	2.08 kg	CEREGE, MM	Met. Bull., No. 102, MAPS 48, in preparation (2014)
Catalina 021	Stone, ordinary chondrite	H_3	W_2		12.6 ± 7.4	11.8 ± 4.7	0.8 ± 0.6		5.06	2009	320 g	CEREGE	Met. Bull., No. 102, MAPS 48, in preparation (2014)
Catalina 022	Stone, ordinary chondrite	L_3	W_3		21.7 ± 8.8	21.7 ± 6.4	0.9 ± 0.4		4.21	2009	77 g	CEREGE	Met. Bull., No. 102, MAPS 48, in preparation (2014)
Catalina 023	Stone, ordinary chondrite	H_6	W_2						5.03	2009	53.5 g	CEREGE	Met. Bull., No. 102, MAPS 48, in preparation (2014)
Catalina 024	Stone, ordinary chondrite	H_4	W_1		18.2 ± 0.1	16.0 ± 0.2	1.3 ± 0.7		5.39	2009	312 g	CEREGE	Met. Bull., No. 102, MAPS 48, in preparation (2014)
Catalina 025	Stone, ordinary chondrite	L_6	W_1						4.69	2009	39 g	CEREGE	Met. Bull., No. 102, MAPS 48, in preparation (2014)
Catalina 026	Stone, ordinary chondrite	H_5	W_1						5.22	2009	845 g	CEREGE	Met. Bull., No. 102, MAPS 48, in preparation (2014)
Catalina 027	Stone, ordinary chondrite	L_6	W_2		24.5	21.5	1.5		4.50	2010	2.41 kg	CEREGE	Met. Bull., No. 102, MAPS 48, in preparation (2014)
Catalina 028	Stone, ordinary chondrite	H_5	W_2		19.2 ± 0.0	17.0 ± 0.1	1.3 ± 0.1		5.36	2010	4.99 kg	CEREGE	Met. Bull., No. 102, MAPS 48, in preparation (2014)

(continued)

Table 5.2 (continued)

Name	Group	Class	W	S	Fa	Fs	Wo	Ni	χ	Found	Weight	Curator	References
Catalina 029	Stone, ordinary chondrite	H_5	W_2			17.8	1.8		4.75	2010	169 g	CEREGE	Met. Bull., No. 102, MAPS 48, in preparation (2014)
Catalina 030	Stone, ordinary chondrite	H_5	W_2						5.01	2010	214 g	CEREGE	Met. Bull., No. 102, MAPS 48, in preparation (2014)
Catalina 031	Stone, ordinary chondrite	L_6	W_2						4.59	2010	1,178 g	CEREGE	Met. Bull., No. 102, MAPS 48, in preparation (2014)
Catalina 032	Stone, ordinary chondrite	H_4	W_2		19.4 ± 0.7	17.2 ± 0.3	1.4 ± 0.9		5.08	2010	1,107 g	CEREGE	Met. Bull., No. 102, MAPS 48, in preparation (2014)
Catalina 033	Stone, ordinary chondrite	L_6	W_3						4.25	2010	211 g	CEREGE	Met. Bull., No. 102, MAPS 48, in preparation (2014)
Catalina 034	Stone, ordinary chondrite	LL_5	W_4		26.8	22.5	1.3		4.05	2010	20 g	CEREGE	Met. Bull., No. 102, MAPS 48, in preparation (2014)
Catalina 035	Stone, ordinary chondrite	H_5	W_1						5.17	2011	904 g	CEREGE, MM	Met. Bull., No. 102, MAPS 48, in preparation (2014)
Catalina 036	Stone, ordinary chondrite	H_5	W_1						5.22	2011	42 g	CEREGE, MM	Met. Bull., No. 102, MAPS 48, in preparation (2014)
Catalina 037	Achondrite, ureilite		Med	S_1	19.9 ± 0.3	16.6 ± 0.2	10.9 ± 0.1			2010	2.22 kg	ASU	Met. Bull., No. 102, MAPS 48, in preparation (2014)

Centre de Recherche et d'Enseignement de Géosciences de l'Environnement (CEREGE); Museo del Meteorito (MM)

Fig. 5.6 Catalina. *Credit*: Carruncho

Fig. 5.7 Catalina 037. *Credit*: Eric Christensen

Fig. 5.8 Cerro del Inca. *Credit*: Matteo Chinellato

Fig. 5.9 Cobija. *Credit*: Jay Piatek

Fig. 5.10 Copiapó. *Credit*: Sergey Vasiliev

Fig. 5.11 Dolores. *Credit*: meteorites.cl

Table 5.3 El Médano

Name	Group	Class	W	Fa	Fo	Wo	χ	Found	Weight	Curator	References
El Médano 001	Stone, ordinary chondrite	H_6	W_2	18.55 ± 0.63	16.45 ± 0.26	1.25 ± 0.09	4.98	2010	10 g	CEREGE, U.Ch	Met. Bull., No. 100, MAPS 46, in preparation (2013)
El Médano 002	Stone, ordinary chondrite	L_6	W_3	25.18 ± 1.28	20.88 ± 0.34	1.33 ± 0.12	4.58	2010	12 g	CEREGE, U.Ch	Met. Bull., No. 100, MAPS 46, in preparation (2013)
El Médano 003	Stone, ordinary chondrite	H_5	W_2	18.57 ± 0.19	16.40 ± 0.16	1.34 ± 0.11	4.99	2010	11 g	CEREGE, U.Ch	Met. Bull., No. 100, MAPS 46, in preparation (2013)
El Médano 004	Stone, ordinary chondrite	H_4	W_1	17.37 ± 0.27	15.15 ± 0.25	1.17 ± 0.76	5.32	2010	151 g	CEREGE, U.Ch	Met. Bull., No. 100, MAPS 46, in preparation (2013)
El Médano 005	Stone, ordinary chondrite	L_6	W_2	24.59 ± 0.25	20.45 ± 0.22	1.66 ± 0.15	4.59	2010	582 g	CEREGE, U.Ch	Met. Bull., No. 100, MAPS 46, in preparation (2013)
El Médano 006	Stone, ordinary chondrite	LL_6	W_4	28.68 ± 0.69	24.22 ± 1.36	2.35 ± 0.31	3.85	2010	55 g	CEREGE, U.Ch	Met. Bull., No. 100, MAPS 46, in preparation (2013)
El Médano 007	Stone, ordinary chondrite	LL_6	W_4	28.34 ± 0.18	23.88 ± 0.18	2.38 ± 0.18	3.46	2010	31 g	CEREGE, U.Ch	Met. Bull., No. 100, MAPS 46, in preparation (2013)
El Médano 008	Stone, ordinary chondrite	L_6	W_3	25.63 ± 1.52	19.99 ± 0.19	1.40 ± 0.24	4.31	2010	33 g	CEREGE, U.Ch	Met. Bull., No. 100, MAPS 46, in preparation (2013)
El Médano 009	Stone, ordinary chondrite	L_6	W_3	24.85 ± 0.92	20.35 ± 0.25	1.52 ± 0.23	4.61	2010	65 g	CEREGE, U.Ch	Met. Bull., No. 100, MAPS 46, in preparation (2013)
El Médano 010	Stone, ordinary chondrite	H_4	W_4	17.29 ± 0.45	15.22 ± 1.11	1.58 ± 0.38	4.57	2010	7 g	CEREGE, U.Ch	Met. Bull., No. 100, MAPS 46, in preparation (2013)

(continued)

Table 5.3 (continued)

Name	Group	Class	W	Fa	Fo	Wo	χ	Found	Weight	Curator	References
El Médano 011	Stone, ordinary chondrite	H_6	W_6	17.92 ± 0.17	16.04 ± 0.43	1.51 ± 0.32	4.81	2010	87 g	CEREGE, U.Ch	Met. Bull., No. 100, MAPS 46, in preparation (2013)
El Médano 012	Stone, ordinary chondrite	H_6	W_3	18.41 ± 0.40	16.10 ± 0.19	1.72 ± 0.02	4.79	2010	54 g	CEREGE, U.Ch	Met. Bull., No. 100, MAPS 46, in preparation (2013)
El Médano 013	Carbonaceous chondrite	CO_3		16.59 ± 16.41	3.03 ± 1.99	3.17 ± 1.19	4.37	2010	5.2 g	CEREGE, U.Ch	Met. Bull., No. 100, MAPS 46, in preparation (2013)
El Médano 014	Stone, ordinary chondrite	H_4	W_2	16.8 ± 0.4	15.1 ± 0.5	0.99 ± 0.4	4.95	2010	54 g	CEREGE, U.Ch	Met. Bull., No. 100, MAPS 46, in preparation (2013)
El Médano 015	Stone, ordinary chondrite	H_4	W_3	17.4 ± 0.6	15.5 ± 0.3	0.9 ± 0.2	4.73	2010	98 g	CEREGE, U.Ch	Met. Bull., No. 100, MAPS 46, in preparation (2013)
El Médano 016	Stone, ordinary chondrite	L_6	W_4	24.7 ± 0.3	20.5 ± 0.3	1.6 ± 0.3	4.01	2010	57 g	CEREGE, U.Ch	Met. Bull., No. 100, MAPS 46, in preparation (2013)
El Médano 017	Stone, ordinary chondrite	L_6	W_3	25.0 ± 0.3	20.3 ± 0.4	1.6 ± 0.2	4.03	2010	156 g	CEREGE, U.Ch	Met. Bull., No. 100, MAPS 46, in preparation (2013)
El Médano 018	Stone, ordinary chondrite	H_4	W_3	16.7 ± 0.2	15.1 ± 0.3	0.7 ± 0.2	4.77	2010	15 g	CEREGE, U.Ch	Met. Bull., No. 100, MAPS 46, in preparation (2013)
El Médano 019	Stone, ordinary chondrite	H_3	W_3	18.30 ± 5.69	1.59 ± 0.80	1.59 ± 0.80	4.48	2010	11 g	CEREGE, U.Ch	Met. Bull., No. 100, MAPS 46, in preparation (2013)
El Médano 020	Stone, ordinary chondrite	L_6	W_3	24.6 ± 0.3	20.5 ± 0.5	1.6 ± 0.1	4.28	2010	102 g	CEREGE, U.Ch	Met. Bull., No. 100, MAPS 46, in preparation (2013)

(continued)

Table 5.3 (continued)

Name	Group	Class	W	Fa	Fo	Wo	χ	Found	Weight	Curator	References
El Médano 021	Stone, ordinary chondrite	H_6	W_3	18.3 ± 0.2	16.2 ± 0.1	1.6 ± 0.1	4.78	2010	50 g	CEREGE, U.Ch	Met. Bull., No. 100, MAPS 46, in preparation (2013)
El Médano 022	Stone, ordinary chondrite	H_6	W_3	18.5 ± 0.3	16.5 ± 0.3	1.7 ± 0.1	4.87	2010	47 g	CEREGE, U.Ch	Met. Bull., No. 100, MAPS 46, in preparation (2013)
El Médano 023	Stone, ordinary chondrite	L_6	W_3	24.0 ± 0.8	20.2 ± 0.4	1.6 ± 0.5	4.37	2010	115 g	CEREGE, U.Ch	Met. Bull., No. 100, MAPS 46, in preparation (2013)
El Médano 024	Stone, ordinary chondrite	H_5	W_3	18.6 ± 0.2	16.8 ± 1.2	1.1 ± 0.2	4.47	2010	9 g	CEREGE, U.Ch	Met. Bull., No. 100, MAPS 46, in preparation (2013)
El Médano 025	Stone, ordinary chondrite	H_6	W_3	18.4 ± 0.2	16.2 ± 0.3	1.4 ± 0.3	4.91	2010	132 g	CEREGE, U.Ch	Met. Bull., No. 100, MAPS 46, in preparation (2013)
El Médano 026	Stone, ordinary chondrite	L_6	W_3	25.2 ± 1.1	20.6 ± 0.4	1.6 ± 0.3	4.71	2010	154 g	CEREGE, U.Ch	Met. Bull., No. 100, MAPS 46, in preparation (2013)
El Médano 027	Stone, ordinary chondrite	H_5	W_3	19.16 ± 0.36	16.81 ± 0.31	1.50 ± 0.13	4.93	2010	46 g	CEREGE, U.Ch	Met. Bull., No. 100, MAPS 46, in preparation (2013)
El Médano 028	Stone, ordinary chondrite	L_5	W_1	24.20 ± 0.37	20.35 ± 0.68	1.35 ± 0.47	4.76	2010	141 g	CEREGE, U.Ch	Met. Bull., No. 100, MAPS 46, in preparation (2013)
El Médano 029	Stone, ordinary chondrite	L_5	W_1	24.00 ± 0.30	20.24 ± 0.15	1.30 ± 0.31	4.90	2010	392 g	CEREGE, U.Ch	Met. Bull., No. 100, MAPS 46, in preparation (2013)
El Médano 030	Stone, ordinary chondrite	H_6	W_3	19.23 ± 0.55	16.43 ± 0.39	1.15 ± 0.07	4.45	2010	32 g	CEREGE, U.Ch	Met. Bull., No. 100, MAPS 46, in preparation (2013)

(continued)

Table 5.3 (continued)

Name	Group	Class	W	Fa	Fo	Wo	χ	Found	Weight	Curator	References
El Médano 031	Stone, ordinary chondrite	H_5	W_2	19.93 ± 0.99	17.65 ± 0.64	1.44 ± 0.14	4.90	2010	35 g	CEREGE, U.Ch	Met. Bull., No. 100, MAPS 46, in preparation (2013)
El Médano 032	Stone, ordinary chondrite	H_6	W_2	18.86 ± 0.12	16.68 ± 0.26	1.74 ± 1.65	4.73	2010	28 g	CEREGE, U.Ch	Met. Bull., No. 100, MAPS 46, in preparation (2013)
El Médano 033	Stone, ordinary chondrite	H_5	W_4	19.02 ± 0.37	16.70 ± 0.34	1.50 ± 0.07	4.48	2010	45 g	CEREGE, U.Ch	Met. Bull., No. 100, MAPS 46, in preparation (2013)
El Médano 034	Stone, ordinary chondrite	H_6	W_3	19.12 ± 0.33	16.45 ± 0.35	1.23 ± 0.19	4.61	2010	10 g	CEREGE, U.Ch	Met. Bull., No. 100, MAPS 46, in preparation (2013)
El Médano 035	Stone, ordinary chondrite	H_5	W_2	19.28 ± 0.54	16.28 ± 0.27	1.08 ± 0.25	4.93	2010	72 g	CEREGE, U.Ch	Met. Bull., No. 100, MAPS 46, in preparation (2013)
El Médano 036	Stone, ordinary chondrite	H_5	W_3	18.45 ± 0.32	16.28 ± 0.21	1.05 ± 0.19	4.77	2010	54 g	CEREGE, U.Ch	Met. Bull., No. 100, MAPS 46, in preparation (2013)
El Médano 037	Stone, ordinary chondrite	L_6	W_3	24.70 ± 0.97	20.38 ± 0.34	1.05 ± 0.22	4.54	2010	111 g	CEREGE, U.Ch	Met. Bull., No. 100, MAPS 46, in preparation (2013)
El Médano 038	Stone, ordinary chondrite	H_6	W_3	18.2 ± 0.2	16.1 ± 0.5	1.4 ± 0.0	4.66	2010	3.8 g	CEREGE, U.Ch	Met. Bull., No. 100, MAPS 46, in preparation (2013)
El Médano 039	Stone, ordinary chondrite	H_5	W_3	16.9 ± 0.2	15.1	1.1	4.64	2010	3.1 g	CEREGE, U.Ch	Met. Bull., No. 100, MAPS 46, in preparation (2013)
El Médano 040	Stone, ordinary chondrite	H_6	W_4	19.3 ± 0.2	16.6	1.7	4.62	2010	3.6 g	CEREGE, U.Ch	Met. Bull., No. 100, MAPS 46, in preparation (2013)

(continued)

Table 5.3 (continued)

Name	Group	Class	W	Fa	Fo	Wo	χ	Found	Weight	Curator	References
El Médano 041	Stone, ordinary chondrite	H_5	W_3	18.4 ± 0.3	15.5	1.2	4.53	2010	1.9 g	CEREGE, U.Ch	Met. Bull., No. 100, MAPS 46, in preparation (2013)
El Médano 042	Stone, ordinary chondrite	L_6	W_3	25.3 ± 1.5	20.5 ± 0.2	1.7 ± 0.1	4.35	2011	22.6 g	CEREGE	Met. Bull., No. 101, MAPS 47, in preparation (2013)
El Médano 043	Stone, ordinary chondrite	H_4	W_2	18.7 ± 0.6	17.2 ± 0.6	1.1 ± 0.1	5.11	2011	24.8 g	CEREGE	Met. Bull., No. 101, MAPS 47, in preparation (2013)
El Médano 044	Stone, ordinary chondrite	L_6	W_3	25.7 ± 0.1	21.7 ± 0.2	1.8 ± 0.3	4.32	2011	44.4 g	CEREGE	Met. Bull., No. 101, MAPS 47, in preparation (2013)
El Médano 045	Stone, ordinary chondrite	H_5	W_2	19.0 ± 0.5	16.8 ± 0.2	1.0 ± 0.1	5.18	2011	319 g	CEREGE	Met. Bull., No. 101, MAPS 47, in preparation (2013)
El Médano 046	Stone, ordinary chondrite	H_4	W_2	17.4 ± 0.2	15.9 ± 0.7	1.2 ± 0.4	4.92	2011	40.1 g	CEREGE	Met. Bull., No. 101, MAPS 47, in preparation (2013)
El Médano 047	Stone, ordinary chondrite	H_4	W_1	18.5 ± 0.1	16.5 ± 0.3	1.2 ± 0.1	5.18	2011	57.9 g	CEREGE	Met. Bull., No. 101, MAPS 47, in preparation (2013)
El Médano 048	Stone, ordinary chondrite	H_5	W_1	19.5 ± 0.5	17.2 ± 0.3	1.0 ± 0.3	5.20	2011	118 g	CEREGE	Met. Bull., No. 101, MAPS 47, in preparation (2013)
El Médano 049	Stone, ordinary chondrite	H_4	W_3	19.0 ± 0.4	17.4 ± 0.1	0.9 ± 0.1	4.81	2011	138 g	CEREGE	Met. Bull., No. 101, MAPS 47, in preparation (2013)
El Médano 050	Stone, ordinary chondrite	H_5	W_4	20.0 ± 0.1	18.1 ± 0.2	1.3 ± 0.2	4.57	2011	52 g	CEREGE	Met. Bull., No. 101, MAPS 47, in preparation (2013)

(continued)

Table 5.3 (continued)

Name	Group	Class	W	Fa	Fo	Wo	χ	Found	Weight	Curator	References
El Médano 051	Stone, ordinary chondrite	L_6	W_2	25.9 ± 0.3	22.2 ± 0.7	1.4 ± 0.2	4.8	2011	28.9 g	CEREGE	Met. Bull., No. 101, MAPS 47, in preparation (2013)
El Médano 052	Stone, ordinary chondrite	H_5	W_2	17.5 ± 0.1	16.0 ± 0.2	1.1 ± 0.0	5.1	2011	66.5 g	CEREGE	Met. Bull., No. 101, MAPS 47, in preparation (2013)
El Médano 053	Stone, ordinary chondrite	$H_{4/5}$	W_2	18.8 ± 0.3	16.7 ± 0.7	1.2 ± 0.0	5.14	2011	17.9 g	CEREGE	Met. Bull., No. 101, MAPS 47, in preparation (2013)
El Médano 054	Stone, ordinary chondrite	H_5	W_2	18.5 ± 0.4	16.4 ± 0.2	1.5 ± 0.2	5.11	2011	46.5 g	CEREGE	Met. Bull., No. 101, MAPS 47, in preparation (2013)
El Médano 055	Stone, ordinary chondrite	H_6	W_2	19.8 ± 0.1	17.5 ± 0.2	1.3 ± 0.1	5.18	2011	34.3 g	CEREGE	Met. Bull., No. 101, MAPS 47, in preparation (2013)
El Médano 056	Carbonaceous chondrite	CK_5		30.0 ± 1.1	24.9 ± 0.8	0.7 ± 0.1	4.51	2011	180 g	CEREGE, MM	Met. Bull., No. 101, MAPS 47, in preparation (2013)
El Médano 057	Stone, ordinary chondrite	H_6	W_4	19.7	17.2	1.2	4.60	2011	10.8 g	CEREGE	Met. Bull., No. 101, MAPS 47, in preparation (2013)
El Médano 058	Stone, ordinary chondrite	H_6	W_2	19.2	16.7	1.0	4.96	2011	11.8 g	CEREGE	Met. Bull., No. 101, MAPS 47, in preparation (2013)
El Médano 059	Stone, ordinary chondrite	H_4	W_3	16.9 ± 0.4	15.4 ± 0.5	0.9 ± 0.2	4.74	2011	27 g	CEREGE	Met. Bull., No. 101, MAPS 47, in preparation (2013)
El Médano 060	Stone, ordinary chondrite	H_3	W_2	17.3 ± 5.1	13.2 ± 4.1	1.8 ± 1.1	4.86	2011	29 g	CEREGE	Met. Bull., No. 101, MAPS 47, in preparation (2013)

(continued)

Table 5.3 (continued)

Name	Group	Class	W	Fa	Fo	Wo	χ	Found	Weight	Curator	References
El Médano 061	Stone, ordinary chondrite	H_5	W_2	18.2 ± 0.1	16.4 ± 0.3	1.8 ± 0.2	5.23	2011	57 g	CEREGE	Met. Bull., No. 101, MAPS 47, in preparation (2013)
El Médano 062	Stone, ordinary chondrite	L_6	W_4	24.7 ± 0.4	21.3 ± 0.3	1.4 ± 0.1	4.12	2011	22 g	CEREGE	Met. Bull., No. 101, MAPS 47, in preparation (2013)
El Médano 063	Stone, ordinary chondrite	H_4	W_2	18.4 ± 0.4	16.2 ± 0.5	1.2 ± 0.3	4.83	2011	23 g	CEREGE	Met. Bull., No. 101, MAPS 47, in preparation (2013)
El Médano 064	Stone, ordinary chondrite	H_4	W_4	18.2 ± 0.2	16.4 ± 0.3	1.5 ± 0.2	4.61	2011	22 g	CEREGE	Met. Bull., No. 101, MAPS 47, in preparation (2013)
El Médano 065	Stone, ordinary chondrite	H_5	W_2	19.6 ± 0.5	17.1 ± 0.3	1.4 ± 0.1	4.97	2011	11.5 g	CEREGE	Met. Bull., No. 101, MAPS 47, in preparation (2013)
El Médano 066	Stone, ordinary chondrite	H_5	W_2	18.3 ± 0.5	16.4 ± 0.6	1.3 ± 0.1	4.92	2011	15.8 g	CEREGE	Met. Bull., No. 101, MAPS 47, in preparation (2013)
El Médano 067	Stone, ordinary chondrite	L_6	W_3	24.3 ± 0.3	20.7 ± 0.4	1.6 ± 0.1	4.03	2011	20.4 g	CEREGE	Met. Bull., No. 101, MAPS 47, in preparation (2013)
El Médano 068	Stone, ordinary chondrite	H_4	W_2	17.4 ± 0.2	15.5 ± 0.3	1.2 ± 0.1	5.24	2011	125 g	CEREGE	Met. Bull., No. 101, MAPS 47, in preparation (2013)
El Médano 069	Stone, ordinary chondrite	H_6	W_2	18.5 ± 0.1	16.5 ± 0.3	1.3 ± 0.1	5.01	2011	65 g	CEREGE	Met. Bull., No. 101, MAPS 47, in preparation (2013)
El Médano 070	Stone, ordinary chondrite	L_6	W_3	25.8 ± 0.8	20.8 ± 0.2	1.3 ± 0.2	4.57	2011	51 g	CEREGE	Met. Bull., No. 101, MAPS 47, in preparation (2013)

(continued)

Table 5.3 (continued)

Name	Group	Class	W	Fa	Fo	Wo	χ	Found	Weight	Curator	References
El Médano 071	Stone, ordinary chondrite	H_5	W_2	18.0 ± 0.1	16.1 ± 0.2	1.0 ± 0.4	5.19	2011	163 g	CEREGE	Met. Bull., No. 101, MAPS 47, in preparation (2013)
El Médano 072	Stone, ordinary chondrite	H_4	W_1	17.6 ± 0.2	17.8 ± 1.2	1.2 ± 0.3	5.13	2011	19.9 g	CEREGE	Met. Bull., No. 101, MAPS 47, in preparation (2013)
El Médano 073	Stone, ordinary chondrite	LL_6	W_3	28.8 ± 0.1	23.9 ± 0.2	1.4 ± 0.0	3.92	2011	67 g	CEREGE	Met. Bull., No. 101, MAPS 47, in preparation (2013)
El Médano 074	Stone, ordinary chondrite	H_4	W_2	18.5 ± 0.3	16.8 ± 0.6	1.2 ± 0.1	5.06	2011	42.4 g	CEREGE	Met. Bull., No. 101, MAPS 47, in preparation (2013)
El Médano 075	Stone, ordinary chondrite	L_4	W_3	24.7 ± 0.7	20.7 ± 0.4	1.5 ± 0.2	4.55	2011	186 g	CEREGE	Met. Bull., No. 101, MAPS 47, in preparation (2013)
El Médano 076	Stone, ordinary chondrite	H_5	W_2	18.6 ± 0.3	16.9 ± 1.1	1.5 ± 0.2	5.25	2011	14.1 g	CEREGE	Met. Bull., No. 101, MAPS 47, in preparation (2013)
El Médano 077	Stone, ordinary chondrite	L_4	W_2	24.2 ± 0.5	16.6 ± 4.4	1.4 ± 1.6	4.72	2011	72 g	CEREGE	Met. Bull., No. 101, MAPS 47, in preparation (2013)
El Médano 078	Stone, ordinary chondrite	L_6	W_3	24.4 ± 0.4	22.2 ± 0.7	1.3 ± 0.1	4.47	2011	1,228 g	CEREGE	Met. Bull., No. 101, MAPS 47, in preparation (2013)
El Médano 079	Stone, ordinary chondrite	H_5	W_2	18.5 ± 0.3	16.3 ± 0.3	1.6 ± 0.1	5.13	2011	230 g	CEREGE	Met. Bull., No. 101, MAPS 47, in preparation (2013)
El Médano 080	Stone, ordinary chondrite	L_6	W_3	25.6 ± 0.5	20.7 ± 0.3	1.4 ± 0.4	4.36	2011	24.5 g	CEREGE	Met. Bull., No. 102, MAPS 48, in preparation (2014)

(continued)

Table 5.3 (continued)

Name	Group	Class	W	Fa	Fo	Wo	χ	Found	Weight	Curator	References
El Médano 081	Stone, ordinary chondrite	H_4	W_3	18.9 ± 0.3	16.7 ± 0.5	1.1 ± 0.0	4.96	2011	13.4 g	CEREGE	Met. Bull., No. 102, MAPS 48, in preparation (2014)
El Médano 082	Stone, ordinary chondrite	H_5	W_2	19.4 ± 0.5	17.2 ± 0.5	1.4 ± 0.1	5.25	2011	22.5 g	CEREGE	Met. Bull., No. 102, MAPS 48, in preparation (2014)
El Médano 083	Stone, ordinary chondrite	H_5	W_1	19.1 ± 0.5	16.7 ± 0.3	1.1 ± 0.1	5.09	2011	12.6 g	CEREGE	Met. Bull., No. 102, MAPS 48, in preparation (2014)
El Médano 084	Stone, ordinary chondrite	H_5	$W_{2/3}$	19.1 ± 0.1	17.2 ± 0.3	1.5 ± 0.2	4.93	2011	45 g	CEREGE	Met. Bull., No. 102, MAPS 48, in preparation (2014)
El Médano 085	Stone, ordinary chondrite	H_4	W_3	18.8 ± 0.3	16.8 ± 0.2	1.0 ± 0.2	4.37	2011	12.9 g	CEREGE	Met. Bull., No. 102, MAPS 48, in preparation (2014)
El Médano 086	Stone, ordinary chondrite	H_4	W_1	18.4 ± 0.2	16.6 ± 0.3	0.9 ± 0.3	5.12	2011	1,258 g	CEREGE	Met. Bull., No. 102, MAPS 48, in preparation (2014)
El Médano 087	Stone, ordinary chondrite	H_4	W_2	18.8 ± 0.3	17.1 ± 0.4	1.2 ± 0.4	4.90	2011	13.9 g	CEREGE	Met. Bull., No. 102, MAPS 48, in preparation (2014)
El Médano 088	Stone, ordinary chondrite	L_6	W_3	24.9 ± 0.2	21.4 ± 0.4	1.5 ± 0.2	4.37	2011	10.1 g	CEREGE	Met. Bull., No. 102, MAPS 48, in preparation (2014)
El Médano 089	Stone, ordinary chondrite	L_6	W_4	24.7 ± 0.3	20.8 ± 0.2	1.7 ± 0.2	4.25	2011	296 g	CEREGE	Met. Bull., No. 102, MAPS 48, in preparation (2014)
El Médano 090	Stone, ordinary chondrite	L_6	W_2	24.3 ± 0.5	20.7 ± 0.3	1.6 ± 0.2	4.44	2011	18.8 g	CEREGE	Met. Bull., No. 102, MAPS 48, in preparation (2014)

(continued)

Table 5.3 (continued)

Name	Group	Class	W	Fa	Fo	Wo	χ	Found	Weight	Curator	References
El Médano 091	Stone, ordinary chondrite	H_5	W_3	18.7 ± 0.4	16.7 ± 0.5	1.5 ± 0.1	4.55	2011	17.8 g	CEREGE	Met. Bull., No. 102, MAPS 48, in preparation (2014)
El Médano 092	Stone, ordinary chondrite	H_6	W_3	18.3 ± 0.4	16.2 ± 0.1	1.5 ± 0.2	4.84	2011	289 g	CEREGE	Met. Bull., No. 102, MAPS 48, in preparation (2014)
El Médano 093	Stone, ordinary chondrite	$H_{4/6}$	W_2	19.9 ± 1.1	16.4 ± 1.6	1.7 ± 0.6	4.95	2011	51.8 g	CEREGE	Met. Bull., No. 102, MAPS 48, in preparation (2014)
El Médano 094	Stone, ordinary chondrite	L_6	W_1	24.5 ± 0.6	20.8 ± 0.6	1.5 ± 0.3	4.74	2011	35.3 g	CEREGE	Met. Bull., No. 102, MAPS 48, in preparation (2014)
El Médano 095	Stone, ordinary chondrite	LL_6	W_3	27.9 ± 0.4	23.0 ± 0.2	1.8 ± 0.1	3.88	2011	72 g	CEREGE	Met. Bull., No. 102, MAPS 48, in preparation (2014)
El Médano 096	Achondrite, acapulcoite		mod	11.6 ± 0.2	11.7 ± 0.3	3.2 ± 0.3	5.44	2011	11.1 g	CEREGE	Met. Bull., No. 102, MAPS 48, in preparation (2014)
El Médano 097	Stone, ordinary chondrite	$L_{5/6}$	$W_{2/4}$	24.9 ± 0.4	21.0 ± 0.6	1.4 ± 0.3	4.32	2011	229 g	CEREGE	Met. Bull., No. 102, MAPS 48, in preparation (2014)
El Médano 098	Stone, ordinary chondrite	L_6	W_4	24.7 ± 0.2	21.1 ± 0.14	1.6 ± 0.3	4.18	2011	135 g	CEREGE	Met. Bull., No. 102, MAPS 48, in preparation (2014)
El Médano 099	Stone, ordinary chondrite	H_5	W_3	18.6 ± 0.2	16.5 ± 0.4	1.5 ± 0.1	4.73	2011	10.5 g	CEREGE	Met. Bull., No. 102, MAPS 48, in preparation (2014)
El Médano 100	Carbonaceous chondrite	C_2		7.3 ± 12.1	3.7 ± 2.8	0.9 ± 0.1	3.93	2011	1.8 g	CEREGE, MNHN	Met. Bull., No. 102, MAPS 48, in preparation (2014)

(continued)

Table 5.3 (continued)

Name	Group	Class	W	Fa	Fo	Wo	χ	Found	Weight	Curator	References
El Médano 101	Stone, ordinary chondrite	L_6	W_3	24.5 ± 0.4	20.3 ± 0.2	1.6 ± 0.2	4.64	2011	19.1 g	CEREGE	Met. Bull., No. 102, MAPS 48, in preparation (2014)
El Médano 102	Stone, ordinary chondrite	$H_{4/5}$	W_1	18.8 ± 0.3	16.3 ± 0.3	1.2 ± 0.1	5.17	2011	65 g	CEREGE	Met. Bull., No. 102, MAPS 48, in preparation (2014)
El Médano 103	Stone, ordinary chondrite	LL_6	W_2	29.8 ± 1.2	24.7 ± 0.9	1.7 ± 0.1	4.00	2011	61 g	CEREGE	Met. Bull., No. 102, MAPS 48, in preparation (2014)
El Médano 104	Stone, ordinary chondrite	LL_5	W_3	29.0 ± 0.3	24.9 ± 0.5	1.8 ± 0.0	3.78	2011	27 g	CEREGE	Met. Bull., No. 102, MAPS 48, in preparation (2014)
El Médano 105	Stone, ordinary chondrite	$H_{5/6}$	W_3	18.4 ± 0.2	16.4 ± 0.3	1.3 ± 0.2	5.00	2011	16.7 g	CEREGE	Met. Bull., No. 102, MAPS 48, in preparation (2014)
El Médano 106	Stone, ordinary chondrite	$H_{5/6}$	W_2	18.9 ± 0.6	16.4 ± 0.1	1.6 ± 0.2	4.98	2011	91 g	CEREGE	Met. Bull., No. 102, MAPS 48, in preparation (2014)
El Médano 107	Stone, ordinary chondrite	$H_{5/6}$	W_3	18.7 ± 0.3	16.6 ± 0.8	1.3 ± 0.2	4.75	2011	10.3 g	CEREGE	Met. Bull., No. 102, MAPS 48, in preparation (2014)
El Médano 108	Stone, ordinary chondrite	H_4	W_2	18.7 ± 0.2	16.9 ± 0.2	1.2 ± 0.2	4.91	2011	15.5 g	CEREGE	Met. Bull., No. 102, MAPS 48, in preparation (2014)
El Médano 109	Stone, ordinary chondrite	H_3	W_2	17.2 ± 5.7	12.1 ± 5.5	0.8 ± 0.4	5.02	2011	13.1 g	CEREGE	Met. Bull., No. 102, MAPS 48, in preparation (2014)
El Médano 110	Stone, ordinary chondrite	LL_6	W4	32.0 ± 0.3	26.1 ± 1.2	2.2 ± 0.2	3.41	2011	28.8 g	CEREGE	Met. Bull., No. 102, MAPS 48, in preparation (2014)

(continued)

Table 5.3 (continued)

Name	Group	Class	W	Fa	Fo	Wo	χ	Found	Weight	Curator	References
El Médano 111	Stone, ordinary chondrite	H_5	W_4	18.8 ± 0.2	16.9 ± 0.1	1.4 ± 0.2	4.39	2011	139 g	CEREGE	Met. Bull., No. 102, MAPS 48, in preparation (2014)
El Médano 112	Stone, ordinary chondrite	H_5	W_2	18.1 ± 0.3	16.5 ± 0.7	1.8 ± 0.0	5.23	2011	23 g	CEREGE	Met. Bull., No. 102, MAPS 48, in preparation (2014)
El Médano 113	Stone, ordinary chondrite	L_6	W_3	25.1 ± 0.4	21.0 ± 0.3	1.8 ± 0.1	4.05	2011	37 g	CEREGE	Met. Bull., No. 102, MAPS 48, in preparation (2014)
El Médano 114	Stone, ordinary chondrite	H_5	W_3	17.6 ± 0.3	15.9 ± 0.6	1.0 ± 0.2	4.56	2011	24.4 g	CEREGE	Met. Bull., No. 102, MAPS 48, in preparation (2014)
El Médano 115	Stone, ordinary chondrite	H_5	W_1	19.1 ± 0.1	16.3 ± 0.4	1.7 ± 0.4	5.18	2011	84 g	CEREGE	Met. Bull., No. 102, MAPS 48, in preparation (2014)
El Médano 116	Stone, ordinary chondrite	H_6	W_3	18.5 ± 0.2	16.4 ± 0.2	1.4 ± 0.1	4.83	2011	46 g	CEREGE	Met. Bull., No. 102, MAPS 48, in preparation (2014)
El Médano 117	Stone, ordinary chondrite	L_6	W_3	25.6 ± 0.4	21.6 ± 0.1	1.7 ± 0.2	4.68	2011	52 g	CEREGE	Met. Bull., No. 102, MAPS 48, in preparation (2014)
El Médano 118	Stone, ordinary chondrite	H_5	W_3	18.4 ± 0.1	16.5 ± 0.3	1.5 ± 0.2	4.99	2011	161 g	CEREGE	Met. Bull., No. 102, MAPS 48, in preparation (2014)
El Médano 119	Stone, ordinary chondrite	H_5	W_2	18.3 ± 0.5	16.2 ± 0.3	1.7 ± 0.0	5.09	2011	23.3 g	CEREGE	Met. Bull., No. 102, MAPS 48, in preparation (2014)
El Médano 120	Stone, ordinary chondrite	H_5	W_2	19.5 ± 0.7	16.5 ± 0.2	1.5 ± 0.0	4.94	2011	24.2 g	CEREGE	Met. Bull., No. 102, MAPS 48, in preparation (2014)

(continued)

Table 5.3 (continued)

Name	Group	Class	W	Fa	Fo	Wo	χ	Found	Weight	Curator	References
El Médano 121	Stone, ordinary chondrite	H_5	W_3	18.2 ± 0.3	16.1 ± 0.2	1.3 ± 0.3	5.09	2011	333 g	CEREGE	Met. Bull., No. 102, MAPS 48, in preparation (2014)
El Médano 122	Stone, ordinary chondrite	$L_{5/6}$	W_3	25.7 ± 0.3	21.3 ± 0.2	1.5 ± 0.2	4.29	2011	55 g	CEREGE	Met. Bull., No. 102, MAPS 48, in preparation (2014)
El Médano 123	Stone, ordinary chondrite	H_5	W_1	19.5 ± 0.4	17.3 ± 0.2	1.2 ± 0.2	5.21	2011	22.1 g	CEREGE	Met. Bull., No. 102, MAPS 48, in preparation (2014)
El Médano 124	Stone, ordinary chondrite	H_5	W_1	18.2 ± 0.4	16.6 ± 0.4	1.1 ± 0.2	5.23	2011	39 g	CEREGE	Met. Bull., No. 102, MAPS 48, in preparation (2014)
El Médano 125	Stone, ordinary chondrite	L_6	W_3	23.7 ± 0.2	20.3 ± 0.3	1.5 ± 0.2	4.49	2011	93 g	CEREGE	Met. Bull., No. 102, MAPS 48, in preparation (2014)
El Médano 126	Stone, ordinary chondrite	H_5	W_3	18.6 ± 0.3	16.6 ± 0.1	1.7 ± 0.1	4.87	2011	584 g	CEREGE	Met. Bull., No. 102, MAPS 48, in preparation (2014)
El Médano 127	Stone, ordinary chondrite	H_5	W_1	18.3 ± 0.2	16.1 ± 0.2	1.3 ± 0.1	5.20	2011	130 g	CEREGE	Met. Bull., No. 102, MAPS 48, in preparation (2014)
El Médano 128	Stone, ordinary chondrite	L_6	W_2	24.8 ± 0.1	21.1 ± 0.4	1.5 ± 0.2	4.55	2011	556 g	CEREGE	Met. Bull., No. 102, MAPS 48, in preparation (2014)
El Médano 129	Stone, ordinary chondrite	L_6	W_2	25.0 ± 0.4	20.7 ± 0.1	1.7 ± 0.2	4.48	2011	27.8 g	CEREGE	Met. Bull., No. 102, MAPS 48, in preparation (2014)
El Médano 130	Stone, ordinary chondrite	L_6	W_3	24.5 ± 0.3	20.3 ± 0.3	1.7 ± 0.2	4.21	2011	43.6 g	CEREGE	Met. Bull., No. 102, MAPS 48, in preparation (2014)

(continued)

Table 5.3 (continued)

Name	Group	Class	W	Fa	Fo	Wo	χ	Found	Weight	Curator	References
El Médano 131	Stone, ordinary chondrite	H_4	W_3	19.1 ± 0.7	16.8 ± 0.5	1.2 ± 0.1	4.86	2011	106 g	CEREGE	Met. Bull., No. 102, MAPS 48, in preparation (2014)
El Médano 132	Stone, ordinary chondrite	H_4	W_1	18.3 ± 0.3	16.5 ± 0.6	0.9 ± 0.2	5.10	2011	670 g	CEREGE	Met. Bull., No. 102, MAPS 48, in preparation (2014)
El Médano 133	Stone, ordinary chondrite	LL_6	W_1	29.8 ± 1.2	24.7 ± 0.9	1.7 ± 0.1	4.00	2011	397 g	CEREGE	Met. Bull., No. 102, MAPS 48, in preparation (2014)
El Médano 134	Stone, ordinary chondrite	H_5	W_2	17.0 ± 0.0	16.4 ± 0.2	1.4 ± 0.1	5.18	2011	20.5 g	CEREGE	Met. Bull., No. 102, MAPS 48, in preparation (2014)
El Médano 135	Stone, ordinary chondrite	L_6	W_3	24.8 ± 0.1	21.2 ± 0.1	1.5 ± 0.3	4.48	2011	71 g	CEREGE	Met. Bull., No. 102, MAPS 48, in preparation (2014)
El Médano 136	Stone, ordinary chondrite	L_5	W_3	24.3 ± 0.4	20.0 ± 0.5	2.7 ± 0.9	4.28	2011	57 g	CEREGE	Met. Bull., No. 102, MAPS 48, in preparation (2014)
El Médano 137	Stone, ordinary chondrite	H_5	W_2	18.0 ± 0.4	16.1 ± 0.4	1.2 ± 0.1	5.05	2011	37 g	CEREGE	Met. Bull., No. 102, MAPS 48, in preparation (2014)
El Médano 138	Stone, ordinary chondrite	L_6	W_3	24.8 ± 0.2	20.7 ± 0.3	1.6 ± 0.1	4.20	2011	17.9 g	CEREGE	Met. Bull., No. 102, MAPS 48, in preparation (2014)
El Médano 139	Stone, ordinary chondrite	H_5	W_2	18.9 ± 0.3	16.9 ± 0.8	1.5 ± 0.1	5.14	2011	11.3 g	CEREGE	Met. Bull., No. 102, MAPS 48, in preparation (2014)
El Médano 140	Stone, ordinary chondrite	$H_{4/5}$	W_3	18.5 ± 0.4	16.5 ± 0.3	1.3 ± 0.3	4.77	2011	11.5 g	CEREGE	Met. Bull., No. 102, MAPS 48, in preparation (2014)

(continued)

Table 5.3 (continued)

Name	Group	Class	W	Fa	Fo	Wo	χ	Found	Weight	Curator	References
El Médano 141	Stone, ordinary chondrite	L_6	W_3	25.2 ± 0.6	21.7 ± 0.9	1.5 ± 0.1	4.35	2011	752 g	CEREGE	Met. Bull., No. 102, MAPS 48, in preparation (2014)
El Médano 142	Stone, ordinary chondrite	L_6	W_3	24.8 ± 0.3	21.0 ± 0.2	1.6 ± 0.1	4.54	2011	20.5 g	CEREGE	Met. Bull., No. 102, MAPS 48, in preparation (2014)
El Médano 143	Stone, ordinary chondrite	H_4	W_2	18.5 ± 0.3	16.9 ± 0.6	1.1 ± 0.2	5.00	2011	59 g	CEREGE	Met. Bull., No. 102, MAPS 48, in preparation (2014)
El Médano 144	Stone, ordinary chondrite	L_4	W_2	24.9 ± 0.4	20.9 ± 0.9	1.4 ± 0.2	4.70	2011	87 g	CEREGE	Met. Bull., No. 102, MAPS 48, in preparation (2014)
El Médano 145	Stone, ordinary chondrite	H_5	W_2	18.0 ± 0.4	15.9 ± 0.1	1.2 ± 0.0	5.01	2011	20.4 g	CEREGE	Met. Bull., No. 102, MAPS 48, in preparation (2014)
El Médano 146	Stone, ordinary chondrite	H_4	W_4	18.7 ± 0.4	16.8 ± 0.5	0.9 ± 0.2	4.62	2011	10.9 g	CEREGE	Met. Bull., No. 102, MAPS 48, in preparation (2014)
El Médano 147	Stone, ordinary chondrite	L_6	W_3	26.0 ± 1.3	20.7 ± 0.2	1.7 ± 0.2	4.50	2011	19.9 g	CEREGE	Met. Bull., No. 102, MAPS 48, in preparation (2014)
El Médano 148	Stone, ordinary chondrite	H_5	W_2	18.5	15.9	1.7	4.95	2011	298 g	CEREGE	Met. Bull., No. 102, MAPS 48, in preparation (2014)
El Médano 149	Stone, ordinary chondrite	H_5	W_2	18.6 ± 0.4	16.4 ± 0.3	1.5 ± 0.1	5.11	2011	99 g	CEREGE	Met. Bull., No. 102, MAPS 48, in preparation (2014)
El Médano 150	Stone, ordinary chondrite	L_6	W_3	23.4 ± 1.3	20.7 ± 0.3	1.7 ± 0.2	4.60	2011	56 g	CEREGE	Met. Bull., No. 102, MAPS 48, in preparation (2014)

(continued)

Table 5.3 (continued)

Name	Group	Class	W	Fa	Fo	Wo	χ	Found	Weight	Curator	References
El Médano 151	Stone, ordinary chondrite	H_4	W_2	18.8 ± 0.2	17.5 ± 0.8	1.2 ± 0.0	5.09	2011	12.6 g	CEREGE	Met. Bull., No. 102, MAPS 48, in preparation (2014)
El Médano 152	Stone, ordinary chondrite	H_4	W_1	18.7 ± 0.5	16.8 ± 0.7	1.4 ± 0.3	5.23	2011	45 g	CEREGE, MM	Met. Bull., No. 102, MAPS 48, in preparation (2014)
El Médano 153	Stone, ordinary chondrite	H_3	W_2	16.6 ± 4.6	11.2 ± 4.7	0.7 ± 0.4	4.99	2011	6.04 kg	CEREGE, MM	Met. Bull., No. 102, MAPS 48, in preparation (2014)
El Médano 154	Stone, ordinary chondrite	L_6	W_4	24.7	21.3	1.9	4.41	2011	1,047 g	CEREGE, MM	Met. Bull., No. 102, MAPS 48, in preparation (2014)
El Médano 155	Stone, ordinary chondrite	H_6	W_1				5.29	2011	517 g	CEREGE, MM	Met. Bull., No. 102, MAPS 48, in preparation (2014)
El Médano 156	Stone, ordinary chondrite	L_6	W_1	25.3	21.7	1.4	4.44	2011	161 g	CEREGE, MM	Met. Bull., No. 102, MAPS 48, in preparation (2014)
El Médano 157	Stone, ordinary chondrite	H_4	W_3	17.8 ± 0.0	13.4 ± 2.6	0.5 ± 0.4	5.08	2011	620 g	CEREGE	Met. Bull., No. 102, MAPS 48, in preparation (2014)
El Médano 158	Stone, ordinary chondrite	H_4	W_1	18.4 ± 0.2	16.3 ± 0.3	0.9 ± 0.3	5.16	2011	38 g	CEREGE	Met. Bull., No. 102, MAPS 48, in preparation (2014)
El Médano 159	Stone, ordinary chondrite	L_6	W_3	24.4 ± 0.3	21.1 ± 0.3	1.5 ± 0.2	4.32	2011	12.3 g	CEREGE	Met. Bull., No. 102, MAPS 48, in preparation (2014)
El Médano 160	Stone, ordinary chondrite	H_6	W_3	19.1 ± 0.1	16.8 ± 0.1	1.4 ± 0.0	4.79	2011	88 g	CEREGE	Met. Bull., No. 102, MAPS 48, in preparation (2014)

(continued)

Table 5.3 (continued)

Name	Group	Class	W	Fa	Fo	Wo	χ	Found	Weight	Curator	References
El Médano 161	Stone, ordinary chondrite	H_5	W_3	17.9 ± 0.1	16.0 ± 0.1	1.1 ± 0.0	4.85	2011	91 g	CEREGE	Met. Bull., No. 102, MAPS 48, in preparation (2014)
El Médano 162	Stone, ordinary chondrite	L_4	W_1	23.4 ± 0.5	20.1 ± 0.3	1.2 ± 0.1	4.89	2011	147 g	CEREGE	Met. Bull., No. 102, MAPS 48, in preparation (2014)
El Médano 163	Stone, ordinary chondrite	L_6	W_3	24.3 ± 0.3	20.7 ± 0.4	1.6 ± 0.1	4.04	2011	20.4 g	CEREGE	Met. Bull., No. 102, MAPS 48, in preparation (2014)
El Médano 164	Stone, ordinary chondrite	H_5	W_3	18.1 ± 0.1	16.4 ± 0.4	1.6 ± 0.2	5.10	2011	282 g	CEREGE	Met. Bull., No. 102, MAPS 48, in preparation (2014)
El Médano 165	Stone, ordinary chondrite	L_6	W_3	25.1 ± 0.4	21.6 ± 1.0	1.8 ± 0.2	4.50	2011	53 g	CEREGE	Met. Bull., No. 102, MAPS 48, in preparation (2014)
El Médano 166	Stone, ordinary chondrite	L_5	W_2	24.3 ± 0.4	20.7 ± 0.5	1.4 ± 0.2	4.79	2011	42 g	CEREGE	Met. Bull., No. 102, MAPS 48, in preparation (2014)
El Médano 167	Stone, ordinary chondrite	H_4	W_2	18.9 ± 0.3	16.8 ± 0.3	1.3 ± 0.1	5.04	2011	163 g	CEREGE	Met. Bull., No. 102, MAPS 48, in preparation (2014)
El Médano 168	Stone, ordinary chondrite	L_6	W_3	25.0 ± 0.1	21.0 ± 0.2	1.5 ± 0.2	4.60	2011	6.7 g	CEREGE	Met. Bull., No. 102, MAPS 48, in preparation (2014)
El Médano 169	Stone, ordinary chondrite	L_6	W_3	23.6 ± 0.1	21.0 ± 1.2	1.5 ± 0.2	4.33	2011	228 g	CEREGE	Met. Bull., No. 102, MAPS 48, in preparation (2014)
El Médano 170	Stone, ordinary chondrite	L_4	W_2	23.3 ± 0.2	19.9 ± 0.2	1.3 ± 0.2	4.80	2011	3.89 kg	CEREGE	Met. Bull., No. 102, MAPS 48, in preparation (2014)

(continued)

Table 5.3 (continued)

Name	Group	Class	W	Fa	Fo	Wo	χ	Found	Weight	Curator	References
El Médano 171	Stone, ordinary chondrite	$H_{\sim 5}$	W_1				5.15	2011	1,821 g	CEREGE	Met. Bull., No. 102, MAPS 48, in preparation (2014)
El Médano 172	Stone, ordinary chondrite	$H_{\sim 5}$	W_2				5.20	2011	1,097 g	CEREGE	Met. Bull., No. 102, MAPS 48, in preparation (2014)
El Médano 173	Stone, ordinary chondrite	$L_{\sim 6}$	W_2				4.61	2011	448 g	CEREGE	Met. Bull., No. 102, MAPS 48, in preparation (2014)
El Médano 174	Stone, ordinary chondrite	$L_{\sim 6}$	W_1				4.58	2011	434 g	CEREGE	Met. Bull., No. 102, MAPS 48, in preparation (2014)
El Médano 175	Stone, ordinary chondrite	$L_{\sim 6}$	W_2				4.38	2011	147 g	CEREGE	Met. Bull., No. 102, MAPS 48, in preparation (2014)
El Médano 176	Stone, ordinary chondrite	L_6	W_2	24.5	21.9	1.1	4.64	2011	64 g	CEREGE	Met. Bull., No. 102, MAPS 48, in preparation (2014)
El Médano 177	Stone, ordinary chondrite	$H_{\sim 5}$	W_1				5.10	2011	1,198 g	CEREGE	Met. Bull., No. 102, MAPS 48, in preparation (2014)
El Médano 178	Stone, ordinary chondrite	$H_{\sim 5}$	W_2				5.25	2011	5.76 kg	CEREGE	Met. Bull., No. 102, MAPS 48, in preparation (2014)
El Médano 179	Stone, ordinary chondrite	H_3	W_3	17.6 ± 2.5	15.8 ± 1.9	0.9 ± 0.4	4.61	2011	14.9 g	CEREGE	Met. Bull., No. 102, MAPS 48, in preparation (2014)
El Médano 180	Stone, ordinary chondrite	$H_{3\text{-}5}$	W_2	18.1 ± 6.4	11.3 ± 6.4	$0.7/1.2 \pm 0.4$	5.08	2011	166 g	CEREGE	Met. Bull., No. 102, MAPS 48, in preparation (2014)

(continued)

Table 5.3 (continued)

Name	Group	Class	W	Fa	Fo	Wo	χ	Found	Weight	Curator	References
El Médano 181	Stone, ordinary chondrite	L_3	W_2	26.0 ± 2.5	21.3 ± 2.7	1.7 ± 1.0	4.65	2011	33.5 g	CEREGE	Met. Bull., No. 102, MAPS 48, in preparation (2014)
El Médano 182	Stone, ordinary chondrite	H_3	W_1	19.9 ± 2.6	16.7 ± 2.6	1.3 ± 0.3	5.12	2011	89 g	CEREGE	Met. Bull., No. 102, MAPS 48, in preparation (2014)
El Médano 183	Stone, ordinary chondrite	H_5	W_4	18.9 ± 0.5	16.7 ± 0.4	1.2 ± 0.1	4.57	2011	6.9 g	CEREGE	Met. Bull., No. 102, MAPS 48, in preparation (2014)
El Médano 184	Stone, ordinary chondrite	H_5	W_1	19.0 ± 0.5	16.5 ± 0.4	1.4 ± 0.1	5.29	2011	474 g	CEREGE	Met. Bull., No. 102, MAPS 48, in preparation (2014)
El Médano 185	Stone, ordinary chondrite	H_5	W_2	19.1 ± 0.4	17.2 ± 1.1	1.5 ± 0.1	5.01	2011	172 g	CEREGE	Met. Bull., No. 102, MAPS 48, in preparation (2014)
El Médano 186	Stone, ordinary chondrite	H_5	W_2	18.9 ± 0.3	16.9 ± 0.4	0.9 ± 0.4	5.23	2011	219 g	CEREGE	Met. Bull., No. 102, MAPS 48, in preparation (2014)
El Médano 187	Stone, ordinary chondrite	H_5	W_3	18.4 ± 0.2	15.9 ± 0.1	1.0 ± 0.2	4.83	2011	61 g	CEREGE	Met. Bull., No. 102, MAPS 48, in preparation (2014)
El Médano 188	Stone, ordinary chondrite	H_6	W_4	18.5 ± 0.3	16.4 ± 0.3	1.7 ± 0.1	4.51	2011	40 g	CEREGE	Met. Bull., No. 102, MAPS 48, in preparation (2014)
El Médano 189	Stone, ordinary chondrite	H_6	W_2	18.1 ± 0.2	16.0 ± 0.2	0.9 ± 0.1	5.18	2011	46 g	CEREGE	Met. Bull., No. 102, MAPS 48, in preparation (2014)
El Médano 190	Stone, ordinary chondrite	H_5	W_2	18.1 ± 0.3	16.2 ± 0.4	1.5 ± 0.2	4.95	2011	2.22 kg	CEREGE	Met. Bull., No. 102, MAPS 48, in preparation (2014)

(continued)

Table 5.3 (continued)

Name	Group	Class	W	Fa	Fo	Wo	χ	Found	Weight	Curator	References
El Médano 191	Stone, ordinary chondrite	H_5	W_2	17.9 ± 0.3	16.0 ± 0.4	1.6 ± 0.2	5.05	2011	138 g	CEREGE	Met. Bull., No. 102, MAPS 48, in preparation (2014)
El Médano 192	Stone, ordinary chondrite	$H_{5/6}$	W_3	18.7 ± 0.3	16.8 ± 0.3	1.5 ± 0.2	4.99	2011	15.7 g	CEREGE	Met. Bull., No. 102, MAPS 48, in preparation (2014)
El Médano 193	Stone, ordinary chondrite	L_6	W_3	24.3	20.6	1.4	4.42	2011	347 g	CEREGE	Met. Bull., No. 102, MAPS 48, in preparation (2014)
El Médano 194	Stone, ordinary chondrite	H_5	W_1				5.20	2011	9.52 kg	CEREGE	Met. Bull., No. 102, MAPS 48, in preparation (2014)
El Médano 195	Stone, ordinary chondrite	H/L_3	W_1	16.9 ± 6.0	17.2 ± 5.6	1.2 ± 0.7	4.99	2011	25.2 g	CEREGE	Met. Bull., No. 102, MAPS 48, in preparation (2014)
El Médano 196	Stone, ordinary chondrite	LL_3	W_2	18.6 ± 10.7	9.9 ± 7.7	0.8 ± 0.6	4.03	2011	18.8 g	CEREGE	Met. Bull., No. 102, MAPS 48, in preparation (2014)

Centre de Recherche et d'Enseignement de Géosciences de l'Environnement (CEREGE); Universiy of Chile (U.Ch); Museo del Meteorito (MM); Museum National d'Histoire Naturelle (MNHN)

Fig. 5.12 El Médano, the largest depository of meteorites in South America. *Credit*: Fritz Junker

Fig. 5.13 Guanaco. *Credit*: Mike Farmer

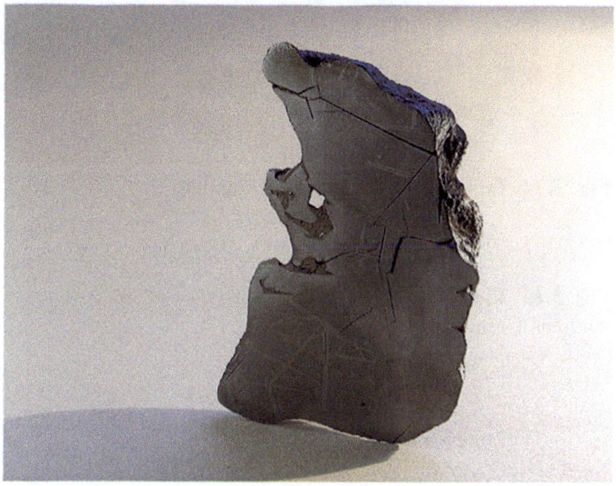

Las Salinas. 23° 0′ S, 69° 30′ W. Antofagasta. Iron, (IIIAB). Find, 1905, 3.52 kg.

Los Vientos. 24° 42′ S, 69° 45′ W. Antofagasta. Finds. 28 meteorites so far. Table 5.4, Fig. 5.20.

Lutschaunig's stone. 27° S, 70° W. Antofagasta. Stone, ordinary chondrite, (L_6). Find, 1861, 100 kg. Natural History Museum.

Mantos Blancos. 23° 27′ S, 70° 7′ W. Antofagasta. Iron, octahedrite, (IVA). Ni = 8.89 wt%. Find, 1876, 10.3 kg. Natural History Museum.

Mantos Blancos 002. 23° 27′ S, 70° 7′ W. Antofagasta. Stone, ordinary chondrite, (L_6), W_3. $Fa_{24.04}$, $Fe_{20.63}$, $Wo_{1.61}$. Find, 2011, 6.8 kg. CEREGE, Museo del Meteorito. Meteoritical Bulletin, no. 102, MAPS 48, in preparation (2014).

María Elena. 22° 20′ S, 69° 40′ W. Antofagasta. Iron, octahedrite (IVA). Ni = 7.64 wt%. Find, 1935, 15.5 kg. Natural History Museum. Fig. 5.21.

Mejillones. 23° 6′ S, 70° 30′ W. Antofagasta. Iron, (IIAB). Find, 1875, 14.83 kg.

Merceditas. 26° 20′ S, 70° 17′ W. Antofagasta. Iron, (IIAB). Find, 1884, 42.9 kg. Natural History Museum.

Monturaqui. 23° 56′ S, 68° 17′ W. Antofagasta. Iron, (IIAB?). Find, 1965, 2 kg. Fig. 5.22.

Fig. 5.14 Imilac. *Credit*: Matteo Chinellato

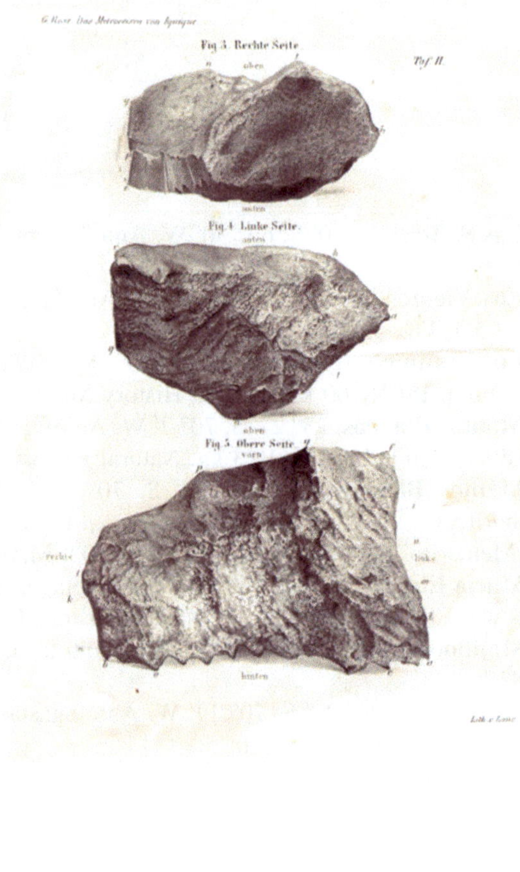

Fig. 5.15 Iquique. *Credit*: Andreas Gren

Fig. 5.16 La Primitiva. *Credit*: Sergey Vasiliev

Fig. 5.17 La Yesera 001. *Credit*: Matteo Chinellato

Fig. 5.18 La Yesera 002. *Credit*: www.meteorites.cl

Fig. 5.19 Las Cruces. *Credit*: Marc Jost (Space Jewels Switzerland Collection) and Thomas Schüpbach

Morro de la Mina. 24° 14′ 48″ S, 68° 51′ 12″ W. Antofagasta. Stone, ordinary chondrite, (H_5). Find, 1986, 1,430 g. University of La Serena, Natural History Museum. Scorzelli et al. (2000).

Negrillos. 19° 53′ S, 69° 50′ W. Tarapacá. Iron, (IIAB). Find, 1936, 28.5 kg. Natural History Museum.

North Chile. 23° S, 69° W. Antofagasta. Iron, (IIAB). Find, 1875, 300 kg. Natural History Museum. Fig. 5.23.

Pampa (a). 23° 12′ S, 70° 26′ W. Antofagasta. Stone, ordinary chondrite, (L6). Find, 1986, 380 g. Natural History Museum. Zolensky et al. (1995). Fig. 5.24.

Pampa (b). 23° 12′ S, 70° 26′ W. Antofagasta. Stone, ordinary chondrite, (L4/5). Find, 1986, 10 kg. Natural History Museum. Fig. 5.25.

Pampa (c). 23° 12′ S, 70° 26′ W. Antofagasta. Stone, ordinary chondrite, (L_4). Find, 1986, 25 kg. Natural History Museum. Fig. 5.26.

Pampa (d). 23° 12′ S, 70° 26′ W. Antofagasta. Stone, ordinary chondrite, (L_5), S_2 $W_{2/3}$. Find, 1986, 12.8 kg. Fig. 5.27.

Pampa (e). 23° 12′ S, 70° 26′ W. Antofagasta. Stone, ordinary chondrite, (L_5), S_1. Find, 1987, 10 kg.

Pampa (f). 23° 11′ S, 70° 26′ W. Antofagasta. Stone, ordinary chondrite, ($L_{4/5}$), S_2 W_2. Find, 2000, 1,300 g. Johnson Space Center. Grossman and Zipfel (2001). Fig. 5.28.

Pampa (g). 23° 11′ S, 70° 26′ W. Antofagasta. Stone, ordinary chondrite, (L_5), S_2 W_3. Find, 2000, 2.9 g. Johnson Space Center. Grossman and Zipfel (2001). Fig. 5.29.

Pampa de Agua Blanca. 24° 10′ S, 69° 50′ W. Antofagasta. Stone, ordinary chondrite, (L_6). Find, 1916, 10 g.

Pampa de Mejillones. Antofagasta. Stone, ordinary chondrites. Meteoritical Bulletin, N° 100, MAPS 46, in preparation (2013). Finds. 14 meteorites so far. Table 5.5, Fig. 5.30.

Table 5.4 Los Vientos

Los Vientos	Group	Class	W	Fa	Fs	Wo	χ	Found	Weight	Curator	References
Lo V 001	Achondrite, diogenite-pm		Minor		24.1–33.4	0.4–3.0		2009	73 g	ASU	Met. Bull., No. 100, MAPS 46, in preparation (2013)
Lo V 002	Stone, ordinary chondrite	L_6	W_1	23.2 ± 0.8	19.8	1.5	4.86	2010	30.75 kg	CEREGE, MM	Met. Bull., No. 100, MAPS 46, in preparation (2013)
Lo V 003	Stone, ordinary chondrite	H_5	W_1	16.9 ± 0.5	14.7	1.1 ± 0.1	5.27	2011	25.45 kg	CEREGE, MM	Met. Bull., No. 100, MAPS 46, in preparation (2013)
Lo V 004	Stone, ordinary chondrite	H_5	W_1	16.4 ± 0.1	14.7 ± 0.1	1.0 ± 0.1	5.30	2011	29.3 kg	CEREGE, MM	Met. Bull., No. 100, MAPS 46, in preparation (2013)
Lo V 005	Stone, ordinary chondrite	H_3	W_1	19.3 ± 11	14.3 ± 6.7	1.3 ± 0.7	5.17	2010	1,431 g	CEREGE	Met. Bull., No. 102, MAPS 48, in preparation (2014)
Lo V 006	Stone, ordinary chondrite	H_4	W_1				5.16	2009	230 g	CEREGE	Met. Bull., No. 102, MAPS 48, in preparation (2014)
Lo V 007	Stone, ordinary chondrite	$H_{5/6}$	W_2	19.3	17.0	1.5	4.95	2010	701 g	CEREGE	Met. Bull., No. 102, MAPS 48, in preparation (2014)
Lo V 008	Stone, ordinary chondrite	H_5	W_1				5.26	2010	1,617 g	CEREGE, MM	Met. Bull., No. 102, MAPS 48, in preparation (2014)
Lo V 009	Stone, ordinary chondrite	H_5	W_1				5.21	2010	85 g	CEREGE, MM	Met. Bull., No. 102, MAPS 48, in preparation (2014)
Lo V 010	Stone, ordinary chondrite	H_5	W_1				5.13	2011	6.8 kg	CEREGE, MM	Met. Bull., No. 102, MAPS 48, in preparation (2014)
Lo V 011	Stone, ordinary chondrite	L_6	W_2				4.51	2011	300 g	CEREGE, MM	Met. Bull., No. 102, MAPS 48, in preparation (2014)
Lo V 012	Stone, ordinary chondrite	$H_{\sim 5}$	W_1				5.30	2011	5.12 kg	CEREGE, MM	Met. Bull., No. 102, MAPS 48, in preparation (2014)
Lo V 013	Stone, ordinary chondrite	H_6	W_1				5.29	2011	2.02 kg	CEREGE, MM	Met. Bull., No. 102, MAPS 48, in preparation (2014)
Lo V 014	Stone, ordinary chondrite	L_6	W_3				4.43	2011	565 g	CEREGE, MM	Met. Bull., No. 102, MAPS 48, in preparation (2014)

(continued)

Table 5.4 (continued)

Los Vientos	Group	Class	W	Fa	Fs	Wo	χ	Found	Weight	Curator	References
Lo V 015	Stone, ordinary chondrite	H_3	W_2	19.0 ± 3.8	13.1 ± 4.6	1.2 ± 0.9	5.03	2009	772 g	CEREGE, MM	Met. Bull., No. 102, MAPS 48, in preparation (2014)
Lo V 016	Stone, ordinary chondrite	LL_3	W_3	13.8 ± 8.7	11.5 ± 6.1	0.7 ± 0.5	4.13	2009	1.9 g	CEREGE, MM	Met. Bull., No. 102, MAPS 48, in preparation (2014)
Lo V 017	Achondrite, ureilite			6.7–16.3	11.4–14.4	4.6–9.3	4.36	2011	109 g	CEREGE, MM	Met. Bull., No. 102, MAPS 48, in preparation (2014)
Lo V 018	Stone, ordinary chondrite	L_6	W_4	25	21.3	1.7	4.13	2010	308 g	CEREGE, MM	Met. Bull., No. 102, MAPS 48, in preparation (2014)
Lo V 019	Stone, ordinary chondrite	H_4	W_1	16.9 ± 0.3	15.2 ± 0.3	0.9 ± 0.1	5.27	2010	263 g	CEREGE, MM	Met. Bull., No. 102, MAPS 48, in preparation (2014)
Lo V 020	Stone, ordinary chondrite	H_4	W_2	17.6 ± 0.1	14.0 ± 1.5	0.8 ± 1.1	5.20	2010	8.1 kg	CEREGE, MM	Met. Bull., No. 102, MAPS 48, in preparation (2014)
Lo V 021	Stone, ordinary chondrite	$L_{\sim 6}$	W_3				4.41	2011	161 g	CEREGE, MM	Met. Bull., No. 102, MAPS 48, in preparation (2014)
Lo V 022	Stone, ordinary chondrite	$L_{\sim 6}$	W_4	24.4			4.47	2011	663 g	CEREGE, MM	Met. Bull., No. 102, MAPS 48, in preparation (2014)
Lo V 023	Stone, ordinary chondrite	$L_{\sim 6}$	W_2	25.2 ± 0.2	21.0	1.3	4.71	2011	7.09 kg	CEREGE, MM	Met. Bull., No. 102, MAPS 48, in preparation (2014)
Lo V 024	Stone, ordinary chondrite	$L_{\sim 6}$	W_2				4.40	2011	333 g	CEREGE, MM	Met. Bull., No. 102, MAPS 48, in preparation (2014)
Lo V 025	Stone, ordinary chondrite	$L_{\sim 6}$	W_2				4.42	2011	506 g	CEREGE, MM	Met. Bull., No. 102, MAPS 48, in preparation (2014)
Lo V 026	Stone, ordinary chondrite	$L_{\sim 6}$	W_2				4.51	2011	203 g	CEREGE, MM	Met. Bull., No. 102, MAPS 48, in preparation (2014)
Lo V 027	Stone, ordinary chondrite	$L_{\sim 6}$	W_3				4.46	2011	131 g	CEREGE, MM	Met. Bull., No. 102, MAPS 48, in preparation (2014)
Lo V 028	Stone, ordinary chondrite	$H_{\sim 5}$	W_2				5.18	2012	12.11 kg	CEREGE, MM	Met. Bull., No. 102, MAPS 48, in preparation (2014)

Arizona State University (ASU); Centre de Recherche et d'Enseignement de Géosciences de l'Environnement (CEREGE); Museo del Meteorito (MM)

Fig. 5.20 Los Vientos 001. *Credit*: Laurence Garvie/ Center for Meteorite Studies/ ASU. Courtesy of the ASU Center for Meteorite Studies

Fig. 5.21 Maria Elena. *Credit*: Mirko Graul

Fig. 5.22 Monturaqui. *Credit*: Jeff Kuyken

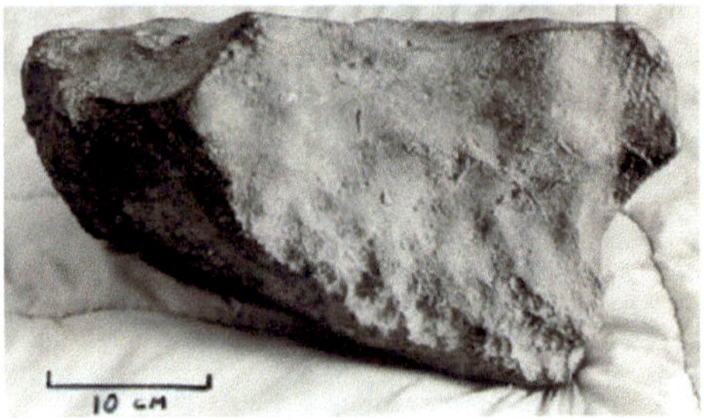

Fig. 5.23 North Chile. *Credit*: Robert Haag

Fig. 5.24 Pampa (a). *Credit*: www.meteorites.cl

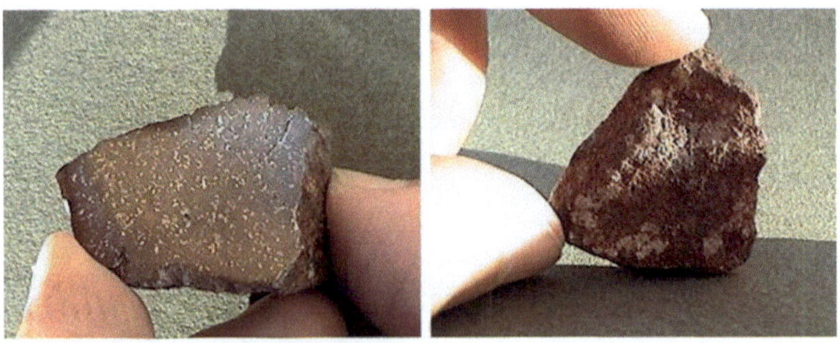

Fig. 5.25 Pampa (b). *Credit*: www.meteorites.cl

Fig. 5.26 Pampa (c). *Credit*: www.meteorites.cl

Fig. 5.27 Pampa (d). *Credit*: www.meteorites.cl

Fig. 5.28 Pampa (f). *Credit*: www.meteorites.cl

Fig. 5.29 Pampa (g). *Credit*: www.meteorites.cl

Pampa Providencia. 24° 27′ S, 69° 34′ 18″ W. Antofagasta. Iron, octahedrite (IIIAB). Ni = 8.86 wt%. Find, 1994, 12.4 kg. University of Arizona. Grossman (1999). Fig. 5.31.

Pan de Azúcar. 26° 30′ S, 69° 30′ W. Atacama. Iron, (IAB complex). Find, 1887, 19.5 kg. Natural History Museum.

Paposo. 25° 0′ S, 70° 28′ W (Paposo 002 to 010: 25° 08′ 36″ S, 70° 19′ 12″ W). Antofagasta. Stone, ordinary chondrite. Finds. 9 meteorites so far. Table 5.6. Figs. 5.32, 5.33 and 5.34.

Pozo Almonte.Unknown coordinates. Iron, octahedrite (IIIAB). Ni = 8.75 wt%. Find, 1990, 7.8 kg. University of La Serena, IGPP-UCLA. Grossman (1998).

Puquios. 27° 9′ S, 69° 55′ W. Atacama. Iron, (IID). Find, 1885, 6.58 kg. Natural History Museum.

Quebrada del León. Unknown coordinates. Stone, ordinary chondrite, (H_6). 1995 Muñoz et al. (2007).

Rencoret 001. Unknown coordinates. Stone, ordinary chondrite, (H_6), S_3 W_3. Find, 1996, 1,992 g. IGPP-UCLA, Museo del Meteorito. Muñoz et al. (2007). Meteoritical Bulletin, no. 101, MAPS 47, in preparation (2013).

Rica Aventura. 21° 59′ S, 69° 37′ W. Antofagasta. Iron, octahedrite, (IVA). Ni = 9.38 wt.%. Find, 1910, 5.4 kg. Olsen and Zeitschel (1979). Fig. 5.35.

Salar de Atacama. 23° 49′ 34″ S, 68° 34′ 20″ W. Antofagasta. Stone, ordinary chondrite, (L_6), S_4 W_3. Find, 2008, 223 g. Museum National d'Histoire Naturelle (France). Meteoritical Bulletin, N° 100, MAPS 46, in preparation (2013).

Salar de Imilac. 24° 12′ 12″ S, 68° 48′ 18″ W. Antofagasta Stone, ordinary chondrite, (H_5), S_3 W_1. Find, 2000, 1,005 g. Arizona State University (USA), Texas Christian University (USA). Grossman and Zipfel (2001). Fig. 5.36.

San Cristóbal. 23° 26′ S, 69° 30′ W. Antofagasta. Iron, (IAB ungrouped). Find, 1882, 5 kg. Natural History Museum.

San Juan. Coordenates see Table 5.7. Antofagasta. Finds. 66 meteorites so far. Figs. 5.37 and 5.38.

Table 5.5 Paposo

Name	Class	S	W	Fa	Fs	Wo	χ	Found	Weight	Curator	References
Paposo	LL_6	S_2	W_2	33	27			2001	2 kg	U. Chile	Russell et al. (2005)
Paposo 002	L/LL_4		W_1	26.5 ± 0.4	18.6 ± 3.3	1.5 ± 1.2	4.37	2011	2.41 kg	CEREGE, MM	Met. Bull., No. 100, MAPS 46, in preparation (2013)
Paposo 003	H_6		W_2	18.8 ± 0.5	16.5 ± 0.1	1.0 ± 0.2	5.44	2011	368 g	CEREGE MM	Met. Bull., No. 101, MAPS 47, in preparation (2013)
Paposo 004	$L_{3.1}$		W_1	21.7 ± 10.8	9.1 ± 5.1	1.2 ± 1.6	4.60	2011	8.25 kg	CEREGE MM	Met. Bull., No. 102, MAPS 48, in preparation (2014)
Paposo 005	H_5		W_2				5.07	2011	1,802 g	CEREGE MM	Met. Bull., No. 102, MAPS 48, in preparation (2014)
Paposo 006	H_3		W_1	15.7 ± 4.4	14.2 ± 4.2	1.3 ± 0.8	5.03	2011	297 g	CEREGE	Met. Bull., No. 102, MAPS 48, in preparation (2014)
Paposo 008	H_5		W_1				5.26	2011	1,035 g	CEREGE MM	Met. Bull., No. 102, MAPS 48, in preparation (2014)
Paposo 009	H_5		W_2				5.21	2011	120 g	CEREGE MM	Met. Bull., No. 102, MAPS 48, in preparation (2014)
Paposo 010	H_6		W_2				5.27	2011	1,569 g	CEREGE MM	Met. Bull., No. 102, MAPS 48, in preparation (2014)

University of Chile (U. Ch); Centre de Recherche et d'Enseignement de Géosciences de l'Environnement (CEREGE); Museo del Meteorito (MM)

Fig. 5.30 Pampa de Mejillones. *Credit*: Manuel Moncada Merino

Fig. 5.31 Pampa Providencia. *Credit*: Matteo Chinellato

Table 5.6 Pampa de Mejillones

Pampa de Mejillones	Coordinates	Class	S	W	Fa	Fs	Wo	χ	Found	Weight	Curator
PdM 001	23° 09′ 23″ S, 70° 28′ 28″ W	L_5		W_4	24	20.1			1999	635 g	NMNH (SI)
PdM 002	23° 12′ 51″ S, 70° 26′ 51″ W	H_5	S_2	W_3	18.3	16.9		4.88	2003	162 g	NMNH (SI)
PdM 003	23° 15′ 48″ S, 70° 27′ 16″ W	H_5	S_2	W_2	19.4	16.9			2003	321 g	NMNH (SI)
PdM 004	23° 12′ 17″ S, 70° 27′ 01″ W	L_6	S_3	$W_{4/5}$	25.6	21.6		4.34	2003	3.16 kg	JSC, U. Chile
PdM 005	23° 13′ 17″ S, 70° 27′ 50″ W	H_4	S_1	W_4	19.1	15.1			2003	222 g	NMNH (SI)
PdM 006	23° 12′ 47″ S, 70° 27′ 16″ W	L_5	S_2	W_2	24.5	20.7			2003	67 g	NMNH (SI)
PdM 007	23° 13′ 45″ S, 70° 27′ 21″ W	L_6	S_3	W_4	24.8	21.2		4.77	2003	1,075 g	NMNH (SI)
PdM 008	23° 11′ 17″ S, 70° 31′ 00″ W	H_5	S_1	W_4	18.7	16.4			2003	25 g	NMNH (SI)
PdM 009	23° 16′ 13″ S, 70° 27′ 30″ W	H_5	S_1	W_4	18.5	16.5			2003	238 g	NMNH (SI)
PdM 010	23° 12′ 08″ S, 70° 26′ 07″ W	L_5	S_3	W_3	25.1	21.8		4.39	2004	360 g	NHM
PdM 011	23° 08′ 10″ S, 70° 29′ 31″ W	L_5	$S_{4/5}$	W_5	25.3	21.5		4.32	2004	46 g	NHM
PdM 012	23° 09′ 44″ S, 70° 26′ 11″ W	H_4	S_2	W_5	18.34 ± 0.10	16.33 ± 0.55	0.99 ± 0.38	4.77	2006	360 g	CEREGE, U. Chile
PdM 013	23° 13′ 24″ S, 70° 26′ 25″ W	H_6	S_2	W_5	20.06 ± 0.57	17.14 ± 0.17	1.56 ± 0.1	4.47	2006	46 g	U. Chile
PdM 014	23° 13′ 49″ S, 70° 25′ 20″ W	L/LL_{4-6}	S_3	W_2	26.42 ± 0.55	22.18 ± 0.99		4.55	2006	3.65 kg	CEREGE, U. Chile

National Museum of Natural History Smithsonian Institution [NMNH (SI)]; Johnson Space Center (JSC); University of Chile (U. Ch); Natural History Museum (NHM); Centre de Recherche et d'Enseignement de Géosciences de l'Environnement (CEREGE)

Fig. 5.32 Paposo. *Credit*: Millarca Valenzuela and Enrique Stucken

Fig. 5.33 Paposo. *Credit*: Google Earth © 2013 Mapcity Image © 2013 DigitalGlobe

Fig. 5.34 Paposo 002 to 010. *Credit*: Google Earth © 2013 Mapcity Image © 2013 DigitalGlobe Image © 2013 TerraMetrics

San Pedro de Quiles. 31° 01′ S, 71° 24′ W. Coquimbo. Stone, ordinary chondrite, (L_6). Seen falling on 1956, 282 g were recovered. Natural History Museum. Grossman (1999).

Fig. 5.35 Rica Aventura.
Credit: Mike Farmer

Fig. 5.36 Salar de Imilac.
Credit: Mike Farmer

Serranía de Varas. 24° 33′ S, 69° 4′ W. Iron, (IVA). Find, 1875, 1,500 g. Natural History Museum.

Sierra Gorda. 22° 54′ S, 69° 21′ W. Antofagasta. Iron, (IIAB). Find, 1898, 26 kg.

Sierra Sandon. 25° 10′ S, 69° 17′ W. Antofagasta. Iron, (IIIAB). Find, 1923, 6.33 kg. Natural History Museum.

Slaghek's Iron. Unknown coordenates. Iron, (IIIAB). Find, 1916, 1,900 g.

Tamarugal. 20° 48′ S, 69° 40′ W (MetSoc) or 20° 11′ S, 69° 44′ W (Muñoz et al. 2007). Tarapaca. Iron, IIIAB (MetSoc) or IVAB (Muñoz et al. 2007). Find, 1903, 320 kg. Natural History Museum. Vilczek and Wänke (1963). Fig. 5.39.

Tambo del Meteorito. 23° 58′ 52″ S, 68° 18′ 47″ W. Antofagasta. Stone, ordinary chondrite, (H_6). Find, 2002, 13.8 g.

Ternera. 27° 20′ S, 69° 48′ W. Atacama. Iron, (IVB). Find, 1891, 1,980 g. Natural History Museum.

Uasara. Antofagasta. Unknown coordenates. Iron, (IIAB). Ni, 59.5 μg/g. Find date unknown, 3.14 kg were obtained. IGPP-UCLA. Meteoritical Bulletin, no. 100, MAPS 46, in preparation (2013).

Table 5.7 San Juan

San Juan	Coordinates	Group	Class	W	S	Fa	Fs	Wo	χ	Found	Weight	Curator	References
SJ 001	25° 34′ 30″ S, 69° 47′ 42″ W	Stone, ordinary chondrite	L_5			24.5	21.6			2001	1,229 g	JSC, NMNH (SI)	Russell et al. (2003), Gattacceca et al. (2011)
SJ 002	25° 34′ 30″ S, 69° 47′ 42″ W	Stone, ordinary chondrite	H_6			19.25	19			2002	345 g	JSC, NMNH (SI)	Russell et al. (2003)
SJ 003	25° 34.53′ S, 69° 47.7′ W	Stone, ordinary chondrite	H_5	W_2	S_3	19.6 ± 0.55	16.8 ± 0.32		5.18	2007	210 g	CEREGE, U. Chile, MNHNP	Weisberg et al. (2009a, b)
SJ 004	25° 34.53′ S, 69° 47.7′ W	Stone, ordinary chondrite	L_4	W_2	S_3	25.6 ± 1.18	19.4 ± 3.96		4.78	2007	229 g	CEREGE, U. Chile, MNHNP	Weisberg et al. (2009a, b)
SJ 005	25° 34.53′ S, 69° 47.7′ W	Stone, ordinary chondrite	H_6	W_3	S_3	19.4 ± 0.1	17.3 ± 0.61		4.5	2007	186.1 g	CEREGE, U. Chile, MNHNP	Weisberg et al. (2009a, b)
SJ 006	25° 34.53′ S, 69° 47.7′ W	Stone, ordinary chondrite	$H_{3.6}$	W_2	S_2	19.0 ± 0.74	13.1 ± 6.42		4.9	2007	242 g	CEREGE, U. Chile, MNHNP	Weisberg et al. (2009a, b)
SJ 007	25° 34.53′ S, 69° 47.7′ W	Stone, ordinary chondrite	H_6	W_2	S_2	20.5 ± 1.11	17.4 ± 0.63		5.14	2007	399 g	CEREGE, U. Chile, MNHNP	Weisberg et al. (2009a, b)
SJ 008	25° 34.53′ S, 69° 47.7′ W	Stone, ordinary chondrite	LL_6	W_3	S_3	30.6 ± 0.65	25.8 ± 0.45		3.18	2007	103.8 g	CEREGE, U. Chile, MNHNP	Weisberg et al. (2009a, b)

(continued)

Table 5.7 (continued)

San Juan	Coordinates	Group	Class	W	S	Fa	Fs	Wo	χ	Found	Weight	Curator	References
SJ 009	25° 34.53′ S, 69° 47.7′ W	Carbonaceous chondrite	CO_3	W_2	S_1	11.5 ± 12.08	14.8 ± 3.82		4.69	2007	45.6 g	CEREGE, U. Chile, MNHNP, UAz	Weisberg et al. (2009a, b)
SJ 010	25° 34.53′ S, 69° 47.7′ W	Stone, ordinary chondrite	$H_{3.8}$	W_3	S_2	19.5 ± 1.07	16.7 ± 2.31		4.64	2007	21.1 g	CEREGE, U. Chile, MNHNP	Weisberg et al. (2009a, b)
SJ 011	25° 34.53′ S, 69° 47.7′ W	Stone, ordinary chondrite	H_4	W_2	S_2	18.9 ± 0.24	16.9 ± 0.44		5.04	2007	33.7 g	CEREGE, U. Chile, MNHNP	Weisberg et al. (2009a, b)
SJ 012	25° 34.53′ S, 69° 47.7′ W	Stone, ordinary chondrite	H_5	W_2	S_2	19.9 ± 0.24	17.5 ± 0.17		5.06	2007	66.8 g	CEREGE, U. Chile, MNHNP	Weisberg et al. (2009a, b)
SJ 013	25° 35′ S, 69° 47′ W	Stone, ordinary chondrite	L_3	W_1	S_3	24.58 ± 1.07	13.56 ± 8.19		4.73	2008	145 g	CEREGE	Weisberg et al. (2010)
SJ 014	25° 35′ S, 69° 47′ W	Stone, ordinary chondrite	L_6	W_3	S_3	24.38 ± 0.20	20.31 ± 0.22		4.38	2008	134 g	CEREGE	Weisberg et al. (2010)
SJ 015	25° 35′ S, 69° 47′ W	Stone, ordinary chondrite	L_6	W_3	S_3	24.62 ± 0.23	20.50 ± 0.28		4.56	2008	349 g	CEREGE	Weisberg et al. (2010)
SJ 016	25° 35′ S, 69° 47′ W	Stone, ordinary chondrite	H_5	W_2	S_2	18.28 ± 0.11	16.25 ± 0.21		5.16	2008	116 g	CEREGE	Weisberg et al. (2010)
SJ 017	25° 35′ S, 69° 47′ W	Stone, ordinary chondrite	H_6	W_3	S_2	18.73 ± 0.15	16.78 ± 0.36		4.91	2008	56.4 g	CEREGE	Weisberg et al. (2010)

(continued)

Table 5.7 (continued)

San Juan	Coordinates	Group	Class	W	S	Fa	Fs	Wo	χ	Found	Weight	Curator	References
SJ 018	25° 35′ S, 69° 47′ W	Stone, ordinary chondrite	L_5	W_1	S_3	2.02 ± 0.47	20.21 ± 0.15		4.82	2008	17.2 g	CEREGE	Weisberg et al. (2010)
SJ 019	25° 35′ S, 69° 47′ W	Stone, ordinary chondrite	L_6	W_2	S_3	24.41 ± 0.28	21.17 ± 0.54		4.65	2008	181 g	CEREGE	Weisberg et al. (2010)
SJ 020	25° 35′ S, 69° 47′ W	Stone, ordinary chondrite	H_5	W_3	S_3	18.16 ± 0.30	16.14 ± 0.24		4.98	2008	151 g	CEREGE	Weisberg et al. (2010)
SJ 021	25° 35′ S, 69° 47′ W	Stone, ordinary chondrite	H_4	W_3	S_1	18.50 ± 0.31	16.19 ± 0.31		4.94	2008	174 g	CEREGE	Weisberg et al. (2010)
SJ 022	25° 35′ S, 69° 47′ W	Stone, ordinary chondrite	L_6	W_1	S_4	23.94 ± 0.49	21.52 ± 1.17		4.43	2008	86.3 g	CEREGE	Weisberg et al. (2010)
SJ 023	25° 35′ S, 69° 47′ W	Stone, ordinary chondrite	H_5	W_2	S_3	18.95 ± 0.24	16.53 ± 0.23		5.23	2008	539 g	CEREGE	Weisberg et al. (2010)
SJ 024	25° 35′ S, 69° 47′ W	Stone, ordinary chondrite	L_6	W_1	S_3	24.14 ± 0.18	2,038 ± 0.33		4.66	2008	15.2 g	CEREGE	Weisberg et al. (2010)
SJ 025	25° 35′ S, 69° 47′ W	Stone, ordinary chondrite	H_5	W_2	S_3	18.27 ± 0.34	15.54 ± 0.25		5.19	2008	419 g	CEREGE	Weisberg et al. (2010)
SJ 026	25° 35′ S, 69° 47′ W	Stone, ordinary chondrite	L_6	W_2	S_1	24.69 ± 0.26	20.64 ± 0.24		4.44	2008	307 g	CEREGE	Weisberg et al. (2010)

(continued)

Table 5.7 (continued)

San Juan	Coordinates	Group	Class	W	S	Fa	Fs	Wo	χ	Found	Weight	Curator	References
SJ 027	25° 35′ S, 69° 47′ W	Stone, ordinary chondrite	H_{3-5}	W_3	S_2				4.98	2008	199 g	CEREGE	Weisberg et al. (2010)
SJ 028	25° 35′ S, 69° 47′ W	Stone, ordinary chondrite	H_5	W_1	S_3	17.87 ± 0.55	15.32 ± 0.54		5.29	2008	151 g	CEREGE	Weisberg et al. (2010)
SJ 029	25° 35′ S, 69° 47′ W	Stone, ordinary chondrite	H_3	W_3	S_1	18.10 ± 9.30	12.50 ± 4.08		4.93	2008	399 g	CEREGE	Weisberg et al. (2010)
SJ 030	25° 35′ S, 69° 47′ W	Stone, ordinary chondrite	H_5	W_1	S_3	18.71 ± 0.37	16.27 ± 0.25		5.24	2008	25.5 g	CEREGE	Weisberg et al. (2010)
SJ 031	25° 35′ S, 69° 47′ W	Stone, ordinary chondrite	L_3	$W_{0/1}$	S_3	24.13 ± 0.96	17.75 ± 2.79		4.98	2008	218 g	CEREGE	Weisberg et al. (2010)
SJ 032	25° 35′ S, 69° 47′ W	Stone, ordinary chondrite	$H_{5/6}$	W_2	S_3	18.91 ± 0.08	16.56 ± 0.23		5.27	2008	28.8 g	CEREGE	Weisberg et al. (2010)
SJ 033	25° 35′ S, 69° 47′ W	Stone, ordinary chondrite	H_6	W_1	S_4	19.3 ± 0.1	17.1 ± 0.2		5.23	2009	357 g	CEREGE, U. Chile	Garvie (2012)
SJ 034	25° 35′ S, 69° 47′ W	Stone, ordinary chondrite	L_6	W_1	S_4	25.0 ± 0.3	21.1 ± 0.2		5.10	2009	814 g	CEREGE, U. Chile	Garvie (2012)
SJ 035	25° 35′ S, 69° 47′ W	Stone, ordinary chondrite	H_5	W_3	S_2	19.1 ± 0.4	16.8 ± 0.6		4.78	2009	10.3 g	CEREGE, U. Chile	Garvie (2012)

(continued)

Table 5.7 (continued)

San Juan	Coordinates	Group	Class	W	S	Fa	Fs	Wo	χ	Found	Weight	Curator	References
SJ 036	25° 35′ S, 69° 47′ W	Stone, ordinary chondrite	L_6	W_3	S_4	25.1 ± 0.1	21.3 ± 0.2		4.41	2009	27.6 g	CEREGE, U. Chile	Garvie (2012)
SJ 037	25° 35′ S, 69° 47′ W	Stone, ordinary chondrite	L_5	W_2	S_1	24.7 ± 0.1	20.8 ± 0.1		4.72	2009	11.4 g	CEREGE, U. Chile	Garvie (2012)
SJ 038	25° 35′ S, 69° 47′ W	Stone, ordinary chondrite	H_5	W_1	S_3	18.8 ± 0.2	17.4 ± 1.3		5.34	2009	460 g	CEREGE, U. Chile	Garvie (2012)
SJ 039	25° 35′ S, 69° 47′ W	Stone, ordinary chondrite	L_6	W_3	S_2	24.7 ± 0.2	20.7 ± 0.3		4.59	2009	38.5 g	CEREGE, U. Chile	Garvie (2012)
SJ 040	25° 35′ S, 69° 47′ W	Stone, ordinary chondrite	$H_{3\text{-}5}$	W_2	S_2	18.81 ± 3.10	17.0 ± 1.1		4.96	2009	34.5 g	CEREGE, U. Chile	Garvie (2012)
SJ 041	25° 35′ S, 69° 47′ W	Stone, ordinary chondrite	H/L_6	W_2	S_4	22.5 ± 0.2	18.3 ± 0.8		4.59	2009	88.1 g	CEREGE, U. Chile	Garvie (2012)
SJ 042	25° 35′ S, 69° 47′ W	Stone, ordinary chondrite	H_3	W_3	S_1	16.3 ± 4.1	16.4 ± 1.2		4.66	2009	13.9 g	CEREGE, U. Chile	Garvie (2012)
SJ 043	25° 35′ S, 69° 47′ W	Stone, ordinary chondrite	H_5	W_1	S_2	18.7 ± 0.2	16.5 ± 0.3		5.25	2009	26.4 g	CEREGE, U. Chile	Garvie (2012)
SJ 044	25° 35′ S, 69° 47′ W	Stone, ordinary chondrite	H_5	W_1	S_3	19.2 ± 0.5	16.7 ± 0.3		5.23	2009	120 g	CEREGE, U. Chile	Garvie (2012)

(continued)

Table 5.7 (continued)

San Juan	Coordinates	Group	Class	W	S	Fa	Fs	Wo	χ	Found	Weight	Curator	References
SJ 045	25° 35′ S, 69° 47′ W	Stone, ordinary chondrite	H_3	W_3	S_1	16.9 ± 8.7	15.2 ± 3.7		4.73	2009	9.9 g	CEREGE, U. Chile	Garvie (2012)
SJ 046	25° 35′ S, 69° 47′ W	Stone, ordinary chondrite	H_5	W_2	S_3	18.9 ± 0.3	16.9 ± 0.2		4.85	2009	57.4 g	CEREGE, U. Chile	Garvie (2012)
SJ 047	25° 35′ S, 69° 47′ W	Stone, ordinary chondrite	H_5	W_1	S_2	18.6 ± 0.3	16.4 ± 0.6		5.17	2009	14.1 g	CEREGE, U. Chile	Garvie (2012)
SJ 048	25° 35′ S, 69° 47′ W	Stone, ordinary chondrite	H_5	W_1	S_2	18.8 ± 0.3	16.7 ± 0.2		5.17	2009	10.6 g	CEREGE, U. Chile	Garvie (2012)
SJ 049	25° 35′ S, 69° 47′ W	Stone, ordinary chondrite	H_5	W_1	S_1	18.9 ± 0.3	16.6 ± 0.2		5.21	2009	41.8 g	CEREGE, U. Chile	Garvie (2012)
SJ 050	25° 35′ S, 69° 47′ W	Stone, ordinary chondrite	H_6	W_1	S_3	19.9 ± 0.2	17.5 ± 0.3		5.23	2009	29.4 g	CEREGE, U. Chile	Garvie (2012)
SJ 051	25° 35′ S, 69° 47′ W	Stone, ordinary chondrite	H_5	W_1	S_1	19.3 ± 0.2	16.9 ± 0.3		5.20	2009	117 g	CEREGE, U. Chile	Garvie (2012)
SJ 052	25° 35′ S, 69° 47′ W	Stone, ordinary chondrite	L_3	W_1	S_3	22.9 ± 4.6	15.1 ± 7.3		4.73	2009	49 g	CEREGE, U. Chile	Garvie (2012)

(continued)

Table 5.7 (continued)

San Juan	Coordinates	Group	Class	W	S	Fa	Fs	Wo	χ	Found	Weight	Curator	References
SJ 053	25° 26′ 24″ S, 69° 53′ 18″ W	Stone, ordinary chondrite	H_5		W_1	18.44 ± 0.17	15.94 ± 0.27	1.37 ± 0.17	5.26	2010	27 g	CEREGE, U. Chile	Met. Bull., No. 100, MAPS 46, in preparation (2013)
SJ 054	25° 26′ 26″ S, 69° 51′ 49″ W	Stone, ordinary chondrite	L_4		W_1	25.5 ± 0.87	20.65 ± 0.36	1.46 ± 0.21	4.61	2010	238 g	CEREGE, U. Chile	Met. Bull., No. 100, MAPS 46, in preparation (2013)
SJ 055	25° 26′ 22″ S, 69° 51′ 49″ W	Stone, ordinary chondrite	H_3		W_2	11.9 ± 8.2	14.88 ± 4.4	1.2 ± 0.4	4.94	2010	168 g	CEREGE, U. Chile	Met. Bull., No. 100, MAPS 46, in preparation (2013)
SJ 056	25° 26′ 20″ S, 69° 51′ 32″ W	Stone, ordinary chondrite	L_5		W_1	24.54 ± 0.65	20.06 ± 0.44	1.37 ± 0.26	4.70	2010	85 g	CEREGE, U. Chile	Met. Bull., No. 100, MAPS 46, in preparation (2013)
SJ 057	25° 26′ 56″ S, 69° 52′ 53″ W	Stone, ordinary chondrite	L_6		W_1	25.08 ± 0.85	20.8 ± 0.45	1.85 ± 0.13	4.71	2010	263 g	CEREGE, U. Chile	Met. Bull., No. 100, MAPS 46, in preparation (2013)

(continued)

Table 5.7 (continued)

San Juan	Coordinates	Group	Class	W	S	Fa	Fs	Wo	χ	Found	Weight	Curator	References
SJ 058	25° 26′ 58″ S, 69° 52′ 53″ W	Stone, ordinary chondrite	LL_6		W_2	30.39 ± 0.27	25.03 ± 0.25	1.87 ± 0.17	3.25	2010	171 g	CEREGE, U. Chile	Met. Bull., No. 100, MAPS 46, in preparation (2013)
SJ 059	25° 26′ 56″ S, 69° 52′ 49″ W	Stone, ordinary chondrite	L_5		W_1	23.94 ± 0.31	19.92 ± 0.32	1.74 ± 0.39	4.72	2010	168 g	CEREGE, U. Chile	Met. Bull., No. 100, MAPS 46, in preparation (2013)
SJ 060	25° 26′ 47″ S, 69° 53′ 09″ W	Stone, ordinary chondrite	H_5		W_1	18.38 ± 0.36	16.44 ± 0.16	1.55 ± 0.15	5.27	2010	74 g	CEREGE, U. Chile	Met. Bull., No. 100, MAPS 46, in preparation (2013)
SJ 061	25° 26′ 26″ S, 69° 51′ 39″ W	Stone, ordinary chondrite	L_5		W_2	24.0 ± 0.1	20.0 ± 0.1	1.6 ± 0.0	4.64	2010	1.3 g	CEREGE, U. Chile	Met. Bull., No. 100, MAPS 46, in preparation (2013)
SJ 062	25° 25′ 31″ S, 69° 41′ 25″ W	Stone, ordinary chondrite	H_5		W_3	18.2 ± 0.3	16.1 ± 0.1	1.3 ± 0.0	4.64	2010	1,125 g	CEREGE, U. Chile	Met. Bull., No. 100, MAPS 46, in preparation (2013)

(continued)

Table 5.7 (continued)

San Juan	Coordinates	Group	Class	W	S	Fa	Fs	Wo	χ	Found	Weight	Curator	References
SJ 063	25° 35′ S, 69° 47 W	Stone, ordinary chondrite	H_5		W_2				5.12	2010	3.38 kg	CEREGE	Met. Bull., No. 100, MAPS 46, in preparation (2013)
SJ 064	25° 35′ S, 69° 47′ W	Stone, ordinary chondrite	L_6		W_3		21.6	1.3	4.47	2010	656 g	CEREGE	Met. Bull., No. 100, MAPS 46, in preparation (2013)
SJ 065	25° 35′ S, 69° 47′ W	Stone, ordinary chondrite	H_4		W_2	18.3 ± 0.3	16.2 ± 0.2	1.2 ± 0.1	5.00	2010	459 g	CEREGE	Met. Bull., No. 100, MAPS 46, in preparation (2013)
SJ 066	25° 35′ S, 69° 47′ W	Stone, ordinary chondrite	L_6		W_2	24.2	20.4	1.6	4.39	2010	441 g	CEREGE	Met. Bull., No. 100, MAPS 46, in preparation (2013)

Johnson Space Center (JSC); National Museum of Natural History Smithsonian Institution [NMNH (SI)]; Centre de Recherche et d'Enseignement de Géosciences de l'Environnement (CEREGE); University of Chile (U. of Chile); Museum National d'Histoire Naturelle (MNHNP); University of Arizona (UAz)

Fig. 5.37 San Juan 001. *Credit*: www.meteorites.cl

Fig. 5.38 San Juan 002. *Credit*: www.meteorites.cl

Fig. 5.39 Tamarugal. *Credit*: Matteo Chinellato.

Fig. 5.40 Vaca Muerta. *Credit*: Jeff Kuyken

Vaca Muerta. 25° 45′ S, 70° 30′ W. Antofagasta. Stony-Iron, mesosiderite. Find, 1861, 3.83 MT. Pedersen et al. (1992), Wasson (1992), Rull et al. (2004), Kuhn (2008). Natural History Museum. Fig. 5.40.

References

Connolly HC Jr, Smith C, Benedix G, Folco L, Righter K, Zipfel J, Yamaguchi A, Chennaoui Aoudjehane H (2008) The Meteoritical Bulletin, No. 93, 2008 March. Meteorit Planet Sci 43(3):571–632

Domeyko I (1862) Mémoire sur les grandes masses d'aérolithes trouvées au désert d'Atacama dans le voisinage de la Sierra del Chaco. Comptes Rendus de l'Academie des Sciences 55:289–310

Domeyko I (1864a) Sobre las grandes masas de aerolitas halladas en el desierto de Atacama cerca de la Sierra del Chaco. Anales de la Universidad de Chile 25:289–310

Domeyko I (1864b) Mémoire concernant les grandes masses d'aérolithes, trouvées dans le désert d'Atacama, dans le voisinage de la Sierra del Chaco. Annales de Mines 5:431–450

Domeyko I (1875) Note sur deux nouvelles météorites du désert d'Atacama et observations sur les météorites qui ont été découvertes jusqu'ici dans cette partie de l'Amérique méridionale. Comptes Rendus de l'Academie des Sciences 81:599–600

Fletcher L (1889) On the meteorites which have been found in the desert of Atacama and its neighborhood. Mineral Mag 8:224–264

Garvie LAJ (2012) The meteoritical bulletin, No. 99, April 2012. MAPS 47(11):E1–E52

Gattacceca J, Valenzuela M, Uehara M, Jull AJT, Giscard M, Rochette P, Braucher R, Suavet C, Gounelle M, Morata D, Munayco P, Bourot-Denise M, Bourles D, Demory F (2011) The densest meteorite collection area in hot deserts: the San Juan meteorite field (Atacama Desert, Chile). Meteorit Planet Sci 46(9):1276–1287

Grossman JN (1998) The meteoritical bulletin, No. 82*, 1998 July. Meteorit Planet Sci 33:A221–A239

Grossman JN (1999) The meteoritical bulletin, No. 83*, 1999 July. Meteorit Planet Sci 34:A169–A186

Grossman JN (2000) The meteoritical bulletin, No. 84*, 2000 August. Meteorit Planet Sci 35:A199–A225

Grossman JN, Zipfel J (2001) The meteoritical bulletin, No. 85*, 2001 September. Meteorit Planet Sci 36:A293–A322

Killgore BM (1997) Imilac strewnfield, Chile, revisited. LPICo 28:725

Kuhn IA, (2008) Microanálise quantitativa por EDS/MEV das fases metálicas dos meteoritos Putinga e Vaca Muerta. Salão de Iniciação Científica (20. 2008 out. 20–24: Porto Alegre, RS). Livro de resumos. Porto Alegre: UFRGS

Muñoz C, Guerra N, Martínez-Frías J, Lunar R, Cerdá J (2007) The Atacama Desert: a preferential arid region for the recovery of meteorites-find location features and strewnfield distribution patterns. J Arid Environ 71:188–200

Olsen E, Zeitschel W (1979) Rica Aventura—a new iron meteorite from Chile. Meteoritics 14:51–53

Pedersen H, García F (1987) New meteorite finds at Imilac. The Messenger (ESO) 47:1–3

Pedersen H, Canut de Bon C, Lindgren H (1992) Vaca Muerta mesosiderite strewnfield. Meteoritics 27:126–135

Philippi RA (1856) Die Sogenannte Wuste Atacama. Petermanns Geog Mitt 52–71

Philippi RA (1860) Viaje al desierto de Atacama 1853–1854. Edit, Halle, p 254

Rull F, Martínez-Frías J, Sansano A, Medina J, Edwards HGM (2004) A comparative micro-Raman study of Nakhla and Vaca Muerta meteorites. Journal of Raman Spectroscopy, 35: 497–503

Russell S, Zipfel J, Folco L, Jones R, Grady McCoy T, Grossman JN (2003) The meteoritical bulletin, No. 87, 2003 July. Meteorit Planet Sci 38(7):A189–A248

Russell S, Folco L, Grady M, Zolensky M, Jones R, Righter K, Zipfel J, Grossman JN (2004) The meteoritical bulletin, No. 88, 2004 July. Meteorit Planet Sci 39(8):A215–A272

Russell S, Zolensky M, Righter K, Folco L, Jones R, Connolly HC Jr, Grady M, Grossman JN (2005) The meteoritical bulletin, No. 89, 2005 September. Meteorit Planet Sci 40(9):A201–A263

Scherer P, Delisle G (1992) Are there high meteorite concentrations in the Atacama Desert/Chile? Meteoritics 27:A285

Scorzelli RB, Azebedo BS, Antonello LL, Poupeau GR, Neumman R, Canut de Bon C (2000) The Morro de la Mina chondrite revisited. Meteorit Planet Sci 35 (Suppl.), 144

Scott ERD, Wasson JT (1976) Chemical classification of iron meteorites-VIII. Groups IC. IIE, IIIF and 97 other irons. Geochim Cosmochim Acta 40:103–108

Vilczek E, Wänke H (1963) Cosmic ray exposure ages and terrestrial ages of stones and iron meteorites derived from Cl^{36} and Ar^{39} measurements. In Radioactive Dating. IAEA, Vienna, Austria, p 381–393

Wasson JT (1992) The rediscovery of the Vaca Muerta strewnfield. Meteoritics 27:125

Wasson JT, Wang J (1986) Nonmagnetic origin of group-IIE iron meteorites. Geochim Cosmochim Acta 50:725–732

Weisberg MK, Smith C, Benedix G, Folco L, Righter K, Zipfel J, Yamaguchi A, Chennaoui Aoudjehane H (2009) The meteoritical bulletin, No. 95, 2007 March. Meteorit Planet Sci 44(3):1–33

Weisberg MK, Smith C, Benedix G, Herd CDK, Righter K, Haack H, Yamaguchi A, Chennaoui Aoudjehane H, Grossman JF (2009) The meteoritical bulletin, No. 96, September 2009. Meteorit Planet Sci 44(9):1355–1397

Weisberg MK, Smith C, Benedix G, Herd CDK, Righter K, Haack H, Yamaguchi A, Chennaoui Aoudjehane H, Grossman JF (2010) The meteoritical bulletin, No. 97. Meteorit Planet Sci 45(3):449–493

Wlotzka F (1991) The meteoritical bulletin, No. 70*. Meteoritics 26:68–69

Wlotzka F (1993) The meteoritical bulletin, No. 75*. Meteoritics 28:692–703

Zolensky ME, Martinez R, Martinez de los Rios E (1995) New L chondrites from the Atacama Desert, Chile. Meteoritics, 30:785–787

Chapter 6
Colombia

For 200 years, the only meteorite from Colombia was the Santa Rosa iron which was reported by the famous Henry Augustus Ward, remembered as the "Museum Builder to America" (Fig. 6.1). He was an amateur rock hunter who came to Colombia and started to admire and protect that wonderful extraterrestrial iron specimen of about one ton.

Two centuries later another new specimen was discovered (Fig. 6.2). This is because forests and jungles make detection of meteorites extremely difficult.

Cali. 03° 24′ 18″ N, 76° 30′ 36″ W. Valle del Cauca. Stone, ordinary chondrite breccia, (H/L_4), S_3, W_0. Olivine and kamacite compositions are intermediate between H and L. Olivine is equilibrated, but low-Ca pyroxene is very heterogenous. Fall, 2007, 478 g. Smithsonian Institution collections, National Museum of Natural History, IGPP-UCLA, US Geological Survey, Reston, Virginia (USA); Institut de Tècniques Energètiques, Universitat Politècnica de Catalunya, Barcelona (Spain). Connolly et al. (2008), Trigo-Rodríguez et al. (2009a, b). Fig. 6.3.

Santa Rosa. 05° 55′ N, 73° W. Boyacá. Iron IC. Find, 1810, 825 kg. Smithsonian Institution collections, Natural History Museum. Ward (1907), Prieto (1936), Ramírez (1949a , b), Ramírez (1950a, b), Buchwald and Wasson (1968), López (1996), Palmer (2000), Plotkin (2004, 2011), Gil and Concha (2006). Fig. 6.4.

R. D. Acevedo et al., *Catalogue of Meteorites from South America*,
SpringerBriefs in Earth System Sciences, DOI: 10.1007/978-3-319-01925-3_6,

Fig. 6.1 Henry Augustus Ward with the Santa Rosa meteorite. *Credit*: Unknown

Fig. 6.2 Colombian meteorites. Modified from © 2013 Google

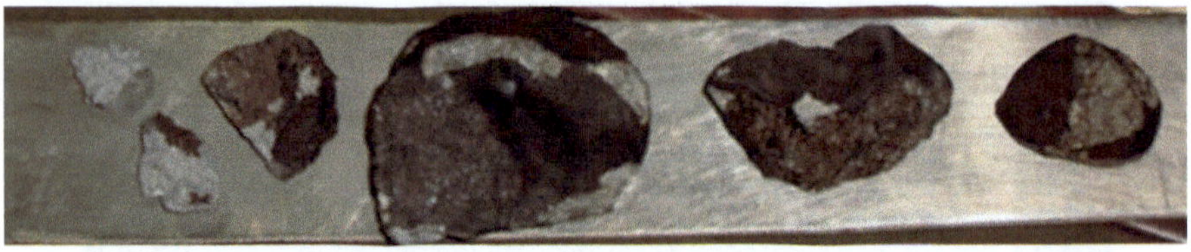

Fig. 6.3 Cali. *Credit*: Mike Farmer

Fig. 6.4 Santa Rosa. *Credit*: Corey Kuo

References

Connolly HC Jr, Smith C, Benedix G, Folco L, Righter K, Zipfel J, Yamaguchi A, Chennaoui Aoudjehane H (2008) The meteoritical bulletin, N° 93, 2008 March. Meteorit Planet Sci 43(3):571–632

Trigo-Rodríguez JM, Llorca J, Rubin AE, Grossman JN, Sears DWG, Naranjo M, Bretzius S, Tapia M, Guarín Sepúlveda MH (2009a) The cali meteorite fall: a new H/L ordinary chondrite. Meteorit Planet Sci 44(2):211–220

Trigo-Rodríguez JM, Llorca J, Sears DWG (2009b) The cali meteorite: luminescence of a recently fallen H/L ordinary chondrite. Conference on Micro-Raman Spectroscopy and Luminescence Studies in the Earth and Planetary Sciences. In Mainz. Germany. LPI Contribution N° 1473:85–86

Ward HA (1907) Colombia meteorite localities: santa rosa, Rasgatá, Tocavita. Am J Sci XXIII, pp. 1–7

Prieto C (1936) Un famoso bólido en Colombia. El de Santa Rosa de Viterbo, Editorial Bodoni

Ramírez JE (1949a) The meteorites of Santa Rosa de Viterbo, Colombia. Popular Astron 57:29

Ramírez JE (1949b) The meteorites of Santa Rosa de Viterbo, Colombia. Contributions of the Meteoritical Society. Popular Astron LVII(1):30–37

Ramírez JE (1950a) La historia del aerolito de Santa Rosa de Viterbo. Boletin de Historia y Antigüedades. Academia Colombiana de Historia, 37(432–434):641–658

Ramírez JE (1950b) Santa Rosa de Viterbo y su famoso aerolito. Juventud Bartolina 28(188):153–155

Buchwald VF, Wasson JT (1968) The two colombian iron meteorites. In: Santa Rosa and Tocavita. Analecta Geológica , Mineralogisk Museum, Kobenhavn, pp 1–19

López F (1996) Cronología del aerolito de santa rosa de viterbo. Colección Museo Nacional de Colombia 874:1–25

Palmer T (2000) Henry augustus ward and his meteorites. Meteorite! 6(2): 39–40

Plotkin H, Henry A (2004) Ward and the recovery of the santa rosa, Meteorite. Meteorit Planet Sci, 39, Supplement. 67th Annual Meeting of the Meteoritical Society, Rio de Janeiro, Colombia, (Brazil), 5038

Plotkin H, Henry A (2011) Ward and the recovery of the santa rosa, Colombia Meteorite. Meteorite

Gil J, Concha A (2006) Caracterización petrográfica y clasificación textural del meteorito de santa rosa de viterbo (Boyacá), Geología Colombiana, Colombia, 31:91–103

Chapter 7
Ecuador

A chondrite remains so far is the only one meteorite Ecuadorian recuperated (Fig. 7.1), and it is a witnessed fall which makes it very special. A noisy explosion was heard along the Daule river on Sunday March 23, 2008, at 09:30 h. After a violent shock wave, four stones were seen to fall separated by 4 km each one. The main stone of 4 kg was found in a farm in a pit of about 1 m in depth and 30 cm in diameter. A second fragment fell on a river at a place called El Pedregal, and the others were lost.

Daule. 52′ 15″ S, 79° 57′ 27″ W. Guayas. Stone, ordinary chondrite, (L_5). Olivine, $Fa_{24.5\pm0.5}$, low-Ca pyroxene, $Fs_{20.8\pm0.5}$ $Wo_{1.4\pm0.2}$. Fall, 2008, ~10.5 kg. Department of Mineral Sciences, National Museum of Natural History, Smithsonian Institution collections, U.S Geological Survey, government agencies of Ecuador and private collectors. Weisberg et al. (2010). Fig. 7.2.

Fig. 7.1 Ecuadorian meteorite. Modified from © 2013 Google

R. D. Acevedo et al., *Catalogue of Meteorites from South America*, 131
SpringerBriefs in Earth System Sciences, DOI: 10.1007/978-3-319-01925-3_7,

Fig. 7.2 Daule. *Credit*: Michael Farmer

Reference

Weisberg MK, Smith C, Benedix G, Herd CDK, Righter K, Haack H, Yamaguchi A, Chennaoui Aoudjehane H, Grossman JF (2010) The meteoritical bulletin, No. 97. Meteorit Planet Sci 45(3): 449–493

Chapter 8
Paraguay

The only fall of a meteorite in this country is the case of the Villarrica (Fig. 8.1). However, there is an interesting but unverified story about a huge metallic body of celestial origin. A very large mass of meteoritic iron was reported to fall during early nineteenth century. Apparently, a very large iron meteorite fell about 500 km from Asuncion. The site was visited by members of the army who estimated its weight in more than 100,000 kg. With great effort they moved the meteorite to the palace of the dictator ruling at that time. Many weapons were made then with its metal. The iron was found to be of a great quality and the story tells that all the mass from that giant meteorite was consumed to make weapons. No sample from this extraterrestrial iron survived the making of new guns. Unfortunately, if truth, this enormous meteorite was totally lost and today it only remains as an unconfirmed record.

R. D. Acevedo et al., *Catalogue of Meteorites from South America*,
SpringerBriefs in Earth System Sciences, DOI: 10.1007/978-3-319-01925-3_8,

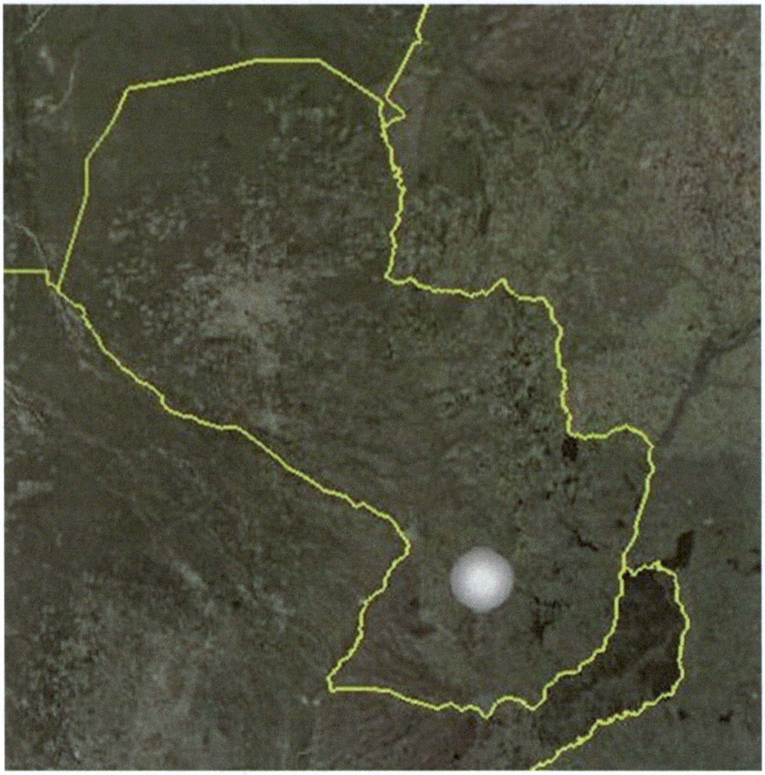

Fig. 8.1 Paraguayan meteorite. Modified from © 2013 Inav/Geosistemas SRL

Villarrica. 25° 50′ S, 56° 30′ W. Guaira. Stone, chondrite. Approved but unclassified. Seen falling, on July 20, 1925 at 19:00 hs. Unknown mass.

Ypacaraí. Pseudometeorite (doubtful stone). Unknown mass. Year fall: 1877.

Chapter 9
Perú

Only two meteorites have been known until at 11:40 local time (16:40 GMT) on September 15, 2007, a chondrite smashed close to the village of Carancas in the Puno Region, near Lake Titicaca on the border with Bolivia (Fig. 9.1). The phenomenon created a small crater larger than 4.5 m (15 ft) deep, and 13 m (44 ft) wide, with visible burned mud ejected all around the impact site. This was a very unusual impact event leaving a small circular structure, showing a very powerful shock wave which hit a large area around the collision site and producing a blast syndrome over people.

Carancas. 16° 39′ 52″ S, 69° 02′ 38″ W. Puno. Stone, ordinary chondrite, ($H_{4/5}$) W_0. Fell on September, 2007, 342 g were collected. Kingsborough Community College of the City University of New York (USA), Planetary and Space Sciences Research Institute, The Open University, Milton Keynes, Lunar and Planetary Laboratory, University of Arizona, Tucson (USA), Connoly et al. (2008), Harris et al. (2008), Núñez Del Prado et al. (2008), Le Pichon et al. (2008), Rosales et al.(2008), Schultz et al. (2008), Tancredi et al. (2008, 2009), Kenkmann et al. (2009), Cerón Loayza and Bravo Cabrejos (2011), Kani et al.(2011), Munayco et al. (2013). Fig. 9.2.

Cerro La Tiza. 14° 31′ 59.56″ S, 75° 46′ 30.40″ W. Ica. Stone, ordinary chondrite, (H_4), S_3 W_3. Olivine $Fa_{17.9\pm0.2}$, pyroxene $Fs_{15.9\pm0.4}$. Find, 2002. The meteorite consists of three pieces with a total mass of 3.74 kg (3.475 kg, 235 g, and 30 g). Fragments are irregularly formed, shiny and of brownish color with very small black patches of fusion crust left. Max-Planck-Institut für Chemie, Abteilung Kosmochemie, Mainz (Germany). Connolly et al. (2007), Schwenzer and Zipfel (2007).

Tambo Quemado. 14° 40′ S, 74° 30′ W. Ayacucho. Iron, (IIIAB). Find, 1950, 141 kg. Natural History Museum. Freyre Villafañe (1950), The Permanent Commission on Meteorites of the International Geological Congress (1958a), Olsen et al. (1993), Cilz and Horejsi (1998). Fig. 9.3.

R. D. Acevedo et al., *Catalogue of Meteorites from South America*,
SpringerBriefs in Earth System Sciences, DOI: 10.1007/978-3-319-01925-3_9,

Fig. 9.1 Peruvian meteorites. Modified from © 2013 Inav/Geosistemas SRL

Fig. 9.2 Carancas. *Credit*: Marcin Cimala

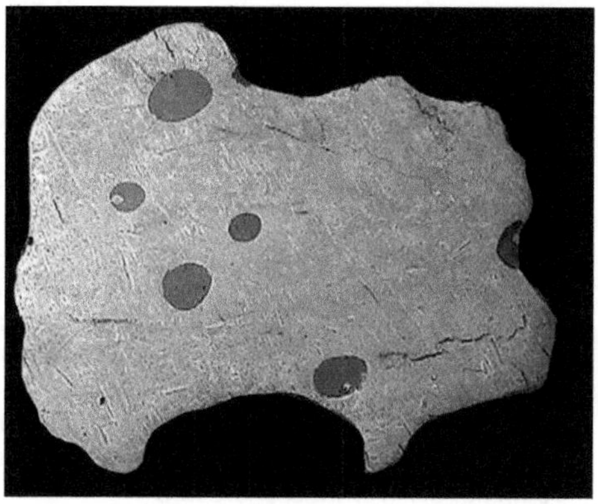

Fig. 9.3 Tambo Quemado. *Credit*: Mike Farmer

References

Connolly HC Jr, Zipfel J, Folco L, Smith C, Jones R, Benedix G, Righter K, Yamaguchi A, Chennaoui Aoudjehane H, Grossman JN (2007) The meteoritical bulletin, No. 91, 2007 March. Meteorit Planet Sci 42(3):413–466

Connolly HC Jr, Smith C, Benedix G, Folco L, Righter K, Zipfel J, Yamaguchi A, Chennaoui Aoudjehane H (2008) The meteoritical bulletin, No. 93, 2008 March. Meteorit Planet Sci 43(3):571–632

Harris RS, Schultz PH, Tancredi G, Ishitsuka J (2008) Preliminary petrologic analysis of impact deformation in the carancas (Perú) cratering event. In: 39th lunar and planetary science conference, (lunar and planetary science XXXIX), LPI Contribution No.1391:2446. League City, Texas

Núñez Del Prado H, Pari W, Ramírez-Cardona M, Macharé J, Macedo L (2008) Reconstruction of an impact event in carancas, Southern Perú, from a GPR study on a small crater. In:Paper presented at 71st annual meeting of the meteoritical society, Meteoritics and Planetary Science Supplement, Matsue 43:5269

Le Pichon A, Antier K, Cansi Y, Hernandez B, Minaya E, Burgoa B, Drob D, Evers LG, Vaubaillon J (2008) Evidence for a meteoritic origin of the september 15, 2007,carancas crater. Meteorit Planet Sci 43(11):1797–1809

Rosales D, Vidal E, Ishitsuka J, Benavente S (2008) Geomagnetic study of carancas meteorite and its crater. In: 39th lunar and planetary science conference, (lunar and planetary science XXXIX), LPI Contribution No. 1391:1744 League City, Texas

Schultz PH, Harris RS, Tancredi G, Ishitsuka J (2008) Implications of the carancas meteorite impact. In: 39th lunar and planetary science conference, (lunar and planetary science XXXIX), LPI Contribution No. 1391:2409 League City, Texas

Schwenzer SP, Zipfel J (2007) Cerro La Tiza, Klassification of the H_4 chondrite in the meteoritical bulletin, 92. Meteorit Planet Sci 42:1663

Tancredi G, Ishitsuka J, Schultz P, Harris RS, Brown P, Revelle D, Antier K, Le Pichon A, Rosales D, Vidal E, Pavel D, Dalmau A, Benavente S, Miranda P, Pereira G, Varela ME, Sánchez L (2008) The carancas crater and meteorite fall: the first recorded impact on earth. Asteroids, Comets, Meteors, LPI Contribution No. 1405: 8260, Baltimore, Maryland

Tancredi G, Ishitsuka J, Schultz PH, Harris RS, Brown P, Revelle DO, Antier K, Le Pichon A, Rosales D, Vidal E, Varela ME, Sánchez L, Benavente S, Bojorquez J, Cabezas D, Dalmau A (2009) A meteorite crater on earth formed on september 15, 2007: the carancas hypervelocity impact. Meteorit Planet Sci 44(12):1967–1984

Kenkmann T, Artemieva NA, Wünnemann K, Poelchau MH, Elbeshausen D, Núñez Del Prado H (2009) The carancas meteorite impact crater, Perú: geologic surveying and modeling of crater formation and atmospheric passage. Meteorit Planet Sci 44(7):985–1000
Cerón Loayza ML, Bravo Cabrejos JA (2011) Characterization of the carancas-puno meteorite by energy dispersive X-ray fluorescence, X-ray diffractometry and transmission Mössbauer spectroscopy. Hyperfine Interact 203(1–3):17–23
Kani R, Hann A, Wickramasinghe C, Di Gregorio BE (2011) Microstructural investigation of the carancas meteorite. Int J Astrobiol 10(2):105–112
Munayco P, Munayco J, Varela ME, Scorzelli RB (2013) The new peruvian meteorite carancas: mössbauer spectroscopy and X-Ray diffraction studies. Earth Moon Planet 110(1–2):1–9
Freyre Villafañe A (1950) Meteorito de Tambo Quemado. Boletín del Instituto Nacional de Investigación y Fomento Mineros, Año 1, no. 1: 141–143, Lima, Perú
The Permanent Commission on Meteorites of the International Geological Congress (1958a) Meteorites not included in the prior-hey catalogue of meteorites 1953, The meteoritical bulletin, 8:1–10, Moscow, URSS
Olsen E, Hutcheon I, Moore C (1993) Tambo quemado: extraordinary concentrations of REE and refractory trace elements caused by artificial heating. In: abstracts of the 24th lunar and planetary science conference, No.1103, Houston, Texas
Cilz M, Horejsi M (1998) The tambo quemado octahedrite from Perú. Meteorite 4:13

Chapter 10
Uruguay

The only meteorite coming from this country could be in fact a meteorite specimen from Argentina. This is the case of the Baygorria iron (Fig. 10.1), which has a very similar composition that of Campo del Cielo's meteorites from Argentina. It could

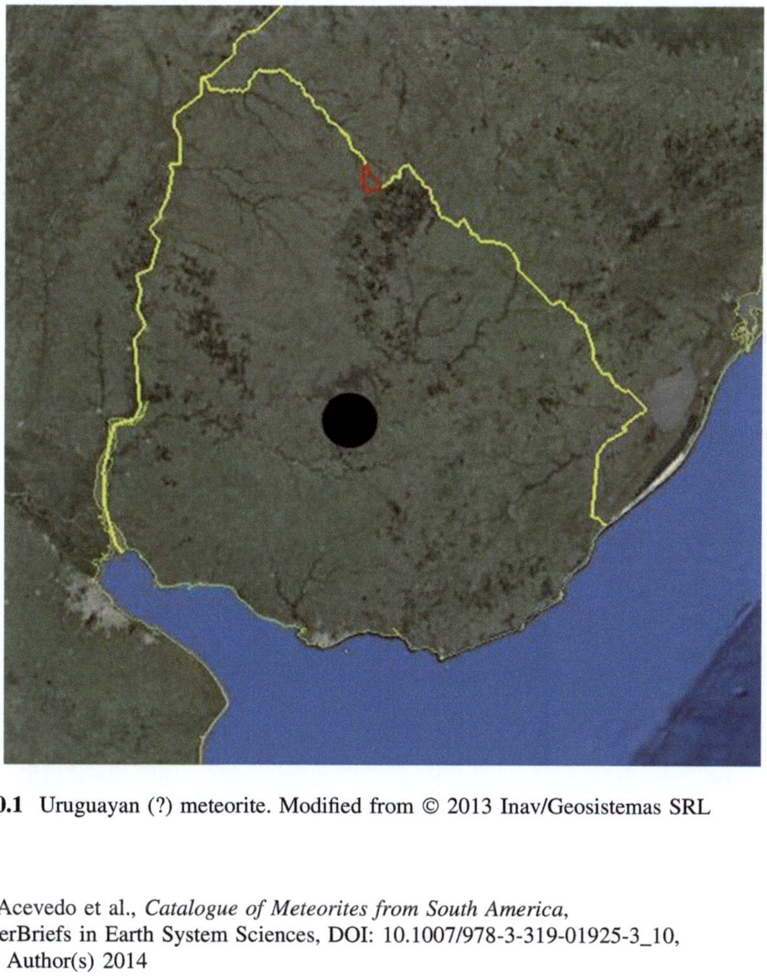

Fig. 10.1 Uruguayan (?) meteorite. Modified from © 2013 Inav/Geosistemas SRL

R. D. Acevedo et al., *Catalogue of Meteorites from South America*, SpringerBriefs in Earth System Sciences, DOI: 10.1007/978-3-319-01925-3_10,

Fig. 10.2 Baygorria (partslice was cut by Marlin Cilz). *Credit*: Rob Wesel

be in fact a meteorite from Campo del Cielo, which was taken by meteorite hunters. Those people could have lie about its precedence to make that iron unique and to get more economic profit of it.

Baygorria. 33° S, 56° W (Fig. 10.2). Provincia de Río Negro. Iron, (IAB). Find, 1994, 80 kg. Smithsonian Institution collections. Grossman (1996). Fake *C.f.* Campo del Cielo meteorite. Fig. 10.2.

Reference

Grossman JN (1996) Catalogs and Inventories. The Meteoritical Bulletin, No. 80, 1996 July. Meteorit Planet Sci 31:A175–A180

Chapter 11
Venezuela

A mass of approximately fifty kilos fell on October 15,1972, killing a cow. This is the case of the Valera meteorite also known as "meteorito asesino" (="murderer meteorite"). This is the only fully documented instance of a meteorite impact causing the death an animal in South America.

There are some other unconfirmed new meteorites like the one from Aragua and the tektites from Anaco.

A map showing the three points where meteorites fell can be seen in Fig. 11.1.

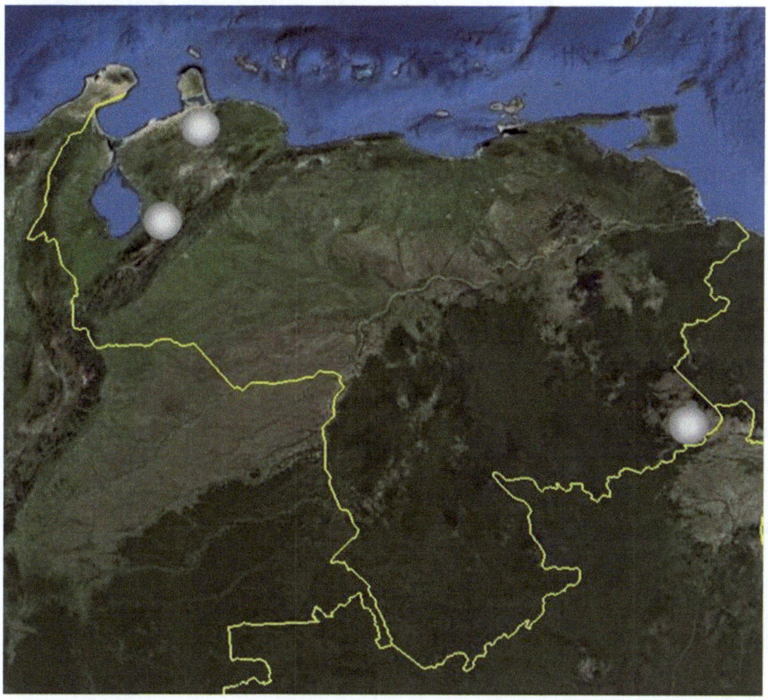

Fig. 11.1 Venezuelan meteorites. Modified from © 2013 Google

R. D. Acevedo et al., *Catalogue of Meteorites from South America*,
SpringerBriefs in Earth System Sciences, DOI: 10.1007/978-3-319-01925-3_11,

Fig. 11.2 Ucera. *Credit*: Franco Urbani

Fig. 11.3 Valera. *Credit*: Unknown

Muenatauray. 04° 54′ N, 61° 12′ W. Bolívar. Iron, (IIAB). Fall, 1960, 30 kg. Lunar and Planetary Laboratory, University of Arizona, Tucson (USA). , Russell et al. (2002).

Ucera. (Synonyms: Coro). 11° 03′ N, 69° 51′ W. Falcon. Stone, olivine-bronzite chondrite, (H_5). Fall, 1970, 4.59 kg. Instituto Venezolano de Investigaciones Científicas, Smithsonian Institution collections, Natural History Museum. The Commission on Meteorites of the International Union of Geological Sciences (1970), Vaz (1970), Clarke (1971). Fig. 11.2.

Valera. 09° 19′ 00″ N, 70° 37′ 42″ W. Trujillo, Venezuela. Stone, ordinary chondrite, (L_5). Fell on October 15, 1972. 50 kg. IGPP-UCLA, University of California, Los Angeles (USA); Museo Nazionale dell'Antartide, Siena (Italy). Grossman and Zipfel (2001), Folco and Rastelli (2002). Fig. 11.3.

References

Clarke RS Jr (1971) The meteoritical bulletin, 50. Meteoritics 6:111–124

Folco L, Rastelli N (2002) The meteorite collection of the Museo Nazionale dell'Antartide in Siena. Terra Antartica 9(2):101–117

Grossman JN, Zipfel J (2001) The meteoritical bulletin, No 85*, 2001 September. Meteorit Planet Sci 36:A293–A322

Russell SS, Zipfel J, Grossman JN, Grady MM (2002) The meteoritical bulletin, No 86. Meteorit Planet Sci 37(suppl):A157–A184

Vaz JE (1970) Mineralogía y composición química del meteorito "Caserío Ucera". Acta Científica Venezolana 21:157–159

Index

R. D. Acevedo et al., *Catalogue of Meteorites from South America*,
SpringerBriefs in Earth System Sciences, DOI: 10.1007/978-3-319-01925-3,

MIX
Papier aus verantwortungsvollen Quellen
Paper from responsible sources
FSC® C105338

If you have any concerns about our products,
you can contact us on
ProductSafety@springernature.com

In case Publisher is established outside the EU,
the EU authorized representative is:
Springer Nature Customer Service Center GmbH
Europaplatz 3, 69115 Heidelberg, Germany

Printed by Libri Plureos GmbH
in Hamburg, Germany